Agroecology: The Ecology of Sustainable Food Systems

Agroecology: The Ecology of Sustainable Food Systems

Milan Collins

SYRAWOOD
PUBLISHING HOUSE

New York

Published by Syrawood Publishing House,
750 Third Avenue, 9th Floor,
New York, NY 10017, USA
www.syrawoodpublishinghouse.com

Agroecology: The Ecology of Sustainable Food Systems
Milan Collins

International Standard Book Number: 978-1-64740-005-7 (Hardback)

Cataloging-in-Publication Data

Agroecology : the ecology of sustainable food systems / Milan Collins.
    p. cm.
Includes bibliographical references and index.
ISBN 978-1-64740-005-7
1. Agricultural ecology. 2. Agriculture--Environmental aspects.
3. Food supply--Environmental aspects. 4. Sustainable agriculture.
5. Organic farming. I. Collins, Milan.
S589.7 .A37 2020
577.55--dc23

TABLE OF CONTENTS

PREFACE

The study of the ecological processes that are applied to agricultural production systems is known as agroecology. It is involved in the study of various agroecosystems. Agroecology explores the methods by which technology can be used in combination with social, natural and human assets. It also studies the diverse properties related to agroecosystems such as equitability, productivity, stability and sustainability. There are different approaches in this field, like inclusive agroecology, indigenous agroecology and agro-population ecology. It is closely related to the fields of organic agriculture and integrated farming. This textbook attempts to understand the multiple branches that fall under the discipline of agroecology and how such concepts have practical applications. It studies, analyses and upholds the pillars of this field and their utmost significance in modern times. Those in search of information to further their knowledge will be greatly assisted by this book.

To facilitate a deeper understanding of the contents of this book a short introduction of every chapter is written below:

Chapter 1- The study of ecological processes which are applied to the agricultural production systems is known as agroecology. Some of the sub-disciplines and techniques used within this field are agroforestry, permaculture and carbon farming. This chapter has been carefully written to provide an easy understanding of these varied sub-disciplines of agroecology.

Chapter 2- Some of the aspects which are dealt with under agro-ecological land resources assessment are agro-ecological zones assessment, agro-climatic suitability classification and agro-edaphic suitability classification. The chapter closely examines these key techniques of agro-ecological land resources assessment to provide an extensive understanding of the subject.

Chapter 3- The functionally and spatially coherent unit of agricultural activity is known as an agricultural ecosystem. The biodiversity within an agricultural ecosystem is known as agrobiodiversity. All the diverse aspects of agricultural ecosystems as well as their management have been carefully analyzed in this chapter.

Chapter 4- There are a variety of ecological methods which can be used to manage pests and weeds such as crop rotation, cover cropping and biological pest control. This chapter has been carefully written to provide an easy understanding of these facets of managing weeds, pests and insects.

Chapter 5- Plant diseases are an integral part of the function of natural ecosystems. They are responsible for the maintenance of plant populations as well as their composition and diversity. The chapter closely examines the key concepts of plant disease ecology and ecological disease management to provide an extensive understanding of the subject.

I would like to share the credit of this book with my editorial team who worked tirelessly on this book. I owe the completion of this book to the never-ending support of my family, who supported me throughout the project.

Milan Collins

Chapter 1

Understanding Agroecology

The study of ecological processes which are applied to the agricultural production systems is known as agroecology. Some of the sub-disciplines and techniques used within this field are agroforestry, permaculture and carbon farming. This chapter has been carefully written to provide an easy understanding of these varied sub-disciplines of agroecology.

Agroecology is the study of ecological processes applied to agricultural production systems. Bringing ecological principles into agroecosystems could suggest novel management approaches that would not be considered otherwise.

There are three typical ways to define agroecology: as a set of farming practices, as a scientific discipline and as a social movement:

- Farming: Agroecological practices are based on ecological inputs and processes, as well as the provision of ecosystem services. Agroecological practices contribute to the different goals of sustainable agriculture: to provide sufficient food for a growing world population, not to be harmful to the environment and natural resources, to limit use of non-renewable energy, and to ensure economic viability for farmers and their communities. Organic farming, diversified crop rotations, biological pest control, extensive agro-pastoral systems and agroforestry are examples of farming method using agroecology.

- Science: Increasingly, scientific disciplines and networks express concern about dwindling, finite resources such as fossil fuels, about (related) issues such as climate change, about soil, biodiversity, health and mores.

Although 'business-as-usual' is no longer an option, much mainstream agricultural thinking is focused on retaining the high input industrialised type of farming exemplified by ideas like 'sustainable intensification'.

As a scientific discipline, agroecology studies are quite holistic: they study agroecosystems through an interdisciplinary lens looking at issues such as productivity, stability, sustainability and equitability. They consider issues related to agronomy,ecology, sociology, economics and politics at all relevant scales from the local level to the global level.

- Social Movements: Many organisations, as well as many loosely networked individuals are working towards an agro-ecological food and farming future. Taken together, these can be seen, broadly speaking, as a social movement trying to make agri-food more resource savvy and thus genuinely sustainable in the longer term, more people and environment focused. These include the 150+ organisations who have signed up to the ARC2020 platform, large international organisations like Friends of the Earth, Slow Food and IFOAM, national organisations and Individual farmers and consumers: all can and do contribute to an agroecological future.

What Agroecology is and what it isn't?

So instead of the conventional, monoculture-based industrial approach which relies on external inputs, we need to develop sustainable, regenerative farming systems that improve the well-being of small-scale farmers, create diversity to make food production resilient to a changing and unpredictable climate, and produce sufficient food whilst enhancing biodiversity. Instead of marginalising sustainable local food producers, we need to put sustainable local food at the centre of our food supply, with small-scale producers feeding local communities, rather than being squeezed by industrial-scale global supply chains.

Agroecological farming is needed to preserve natural resources. This includes recycling nutrients and energy on the farm, rather than using external inputs; integrating crop and livestock farming; diversifying species (and therefore genetic resources); and focusing on the ways in which crops and livestock can mutually benefit each other, rather than on individual species. By using organic matter and improving the soil, farmers can promote better plant growth. This is an agro-ecology knowledge-intensive system, but the knowledge is developed by the farmer through understanding local conditions and experimenting.

Re-connecting farmers and consumers is important to help building vibrant local food economies. The aim is to support local producers, processors and retailers, and build links between consumers, local farmers and local food businesses. This means creating decentralized short supply chains, diversified markets based on solidarity and fair prices, and closer links between producers and consumers locally. Consumers should be able to purchase ecologically-produced food from small-scale producers. Short distance distribution models are also an important aspect for the closure of nutrient cycles, a basic need in agro-ecological farming practices. To return plant nutrients back into the loop, back to the soil, on the right spot, in the right composition and in the right amounts, is a complex issue. This complexity increases significantly over distance, so agroecology promotes closed production loop sand minimised external inputs. In this way local food economies answer the basic need for plant nutrients in agroecological farming practices.

There are a myriad of different systems offering 'local food' and 'short supply chains' in Europe, including farmers' markets, 'farm-gate' sales, box delivery schemes, mobile shops, community supported agriculture, consumer-producer cooperatives and collective catering and canteens. Short supply chains are not just about reducing the number of intermediaries, about putting the consumer and the producer at the heart of deciding what is produced, how it is produced, and how to define the value.

Food distribution through short supply chains in local markets have been shown to increase income for producers, add value and generate greater autonomy for farmers, and to strengthen local economies by supporting more small businesses. This can improve the viability of small farms, reduces the carbon footprint from food distribution, and enhances household food security by giving people on low income access to good food and healthy diets, as well as encouraging stronger producer-consumer relationships.

Local food supply chains also create employment in rural areas and bring farmers into direct contact with consumers, encouraging the circulation of revenue locally, all the while enhancing social cohesion and making it more likely that farmers can stay farming. This helps foster a sense of community in rural areas, improving quality of life. It can also provide a basis for education on sustainability and ethical issues in urban areas.

Some Relevant Areas for Agroecology

Production Methods

Sustainability and diversity of farming systems, including:

- Re-connect crop and animal production in order to close nutrient cycles.

- Recycle biomass, optimise and close nutrient cycles and reduce dependence external inputs.

- Improve soil conditions, in particular improving organic matter content and biological activity of the soil.

- Integrate protection of biodiversity with production of food and promote and conserve the genetic diversity of crops and animals.

- Minimise resource losses by managing the micro-climate, increasing soil cover, water harvesting.

Processing and Distribution in an Agroecology Framework

- Relocalised and regionalised agroecological food systems that allow fair prices, create jobs and reconnect consumers to farmers.

- Foods involving a minimum of industrialised inputs and processes.

- Decentralised and innovative community-led local development that empower people in food production and agroecology (access to land, CSAs, rural development, LEADER).

Participation and Decision-making

- Investigate existing power relations, decision-making processes and opportunities for participation in food systems. Strenghten the role of citizens and consumers in food systems.

- Valorise the diversity of knowledge (local / traditional know-how and practices, common and expert knowledge) in the definition of research problems, the definition of people concerned, and in finding solutions.

- Community-based participatory research and innovation which will facilitate the development of diversified seeds and ecological production and distribution systems (by developing meaningful inter-disciplinary networks, involving a wide range of stakeholders to integrate local and traditional knowledge with formal scientific knowledge).

- Acknowledge the similarities and linkages between agricultural systems in the global North and South. The transition towards sustainable food systems demands integrated and simultaneous solutions in North and South.

Important for Consumers

- Increasing awareness among consumers of the negative impacts of business-as-usual in agri-food, as compared to agroecology.

- Supporting systems that empower consumers, helping them become co-producers of the food they eat, in solidarity with other participants in the agri-food system.

- Encouraging and being a voice for consumers, who are are not just consumers but also food citizens operating on the local, regional, national and international levels.

Regenerative Agriculture

"Regenerative Agriculture" describes farming and grazing practices that, among other benefits, reverse climate change by rebuilding soil organic matter and restoring degraded soil biodiversity – resulting in both carbon drawdown and improving the water cycle.

Specifically, Regenerative Agriculture is a holistic land management practice that leverages the power of photosynthesis in plants to close the carbon cycle, and build soil health, crop resilience and nutrient density. Regenerative agriculture improves soil health, primarily through the practices that increase soil organic matter. This not only aids in increasing soil biota diversity and health, but increases biodiversity both above and below the soil surface, while increasing both water holding capacity and sequestering carbon at greater depths, thus drawing down climate-damaging levels of atmospheric CO_2, and improving soil structure to reverse civilization-threatening human-caused soil loss. Research continues to reveal the damaging effects to soil from tillage, applications of agricultural chemicals and salt based fertilizers, and carbon mining. Regenerative Agriculture reverses this paradigm to build for the future.

Regenerative Agricultural Practices

Practices that:

1. contribute to generating/building soils and soil fertility and health,

2. Increase water percolation, water retention, and clean and safe water runoff,

3. Increase biodiversity and ecosystem health and resiliency,

4. Invert the carbon emissions of our current agriculture to one of remarkably significant carbon sequestration thereby cleansing the atmosphere of legacy levels of CO_2.

Practices include:

1. No-till/minimum tillage: Tillage breaks up (pulverizes) soil aggregation and fungal communities while adding excess O_2 to the soil for increased respiration and CO_2 emission. It can be one of the most degrading agricultural practices, greatly increasing soil erosion and carbon loss. A secondary effect is soil capping and slaking that can plug soil spaces for percolation creating much more water runoff and soil loss. Conversely, no-till/minimum tillage, in conjunction with other regenerative practices, enhances soil aggregation, water infiltration and retention, and carbon sequestration. However, some soils benefit from interim ripping to break apart hardpans, which can increase root zones and yields and have the capacity to increase soil health and carbon sequestration. Certain low level chiseling may have similar positive effects.

2. Soil fertility is increased in regenerative systems biologically through application of cover crops, crop rotations, compost, and animal manures, which restore the plant/soil microbiome to promote liberation, transfer, and cycling of essential soil nutrients. Artificial and synthetic fertilizers have created imbalances in the structure and function of microbial communities in soils, bypassing the natural biological acquisition of nutrients for the plants, creating a dependent agroecosystem

and weaker, less resilient plants. Research has observed that application of synthetic and artificial fertilizers contribute to climate change through:

- The energy costs of production and transportation of the fertilizers,

- Chemical breakdown and migration into water resources and the atmosphere,

- The distortion of soil microbial communities including the diminution of soil methanothrops,

- The accelerated decomposition of soil organic matter.

3. Building biological ecosystem diversity begins with inoculation of soils with composts or compost extracts to restore soil microbial community population, structure and functionality restoring soil system energy (Ccompounds as exudates) through full-time planting of multiple crop intercrop plantings, multispecies cover crops, and borders planted for bee habitat and other beneficial insects. This can include the highly successful push-pull systems. It is critical to change synthetic nutrient dependent monocultures, low-biodiversity and soil degrading practices.

4. Well-managed grazing practices stimulate improved plant growth, increased soil carbon deposits, and overall pasture and grazing land productivity while greatly increasing soil fertility, insect and plant biodiversity, and soil carbon sequestration. These practices not only improve ecological health, but also the health of the animal and human consumer through improved micro-nutrients availability and better dietary omega balances. Feed lots and confined animal feeding systems contribute dramatically to:

- Unhealthy monoculture production systems,

- Low nutrient density forage,

- Increased water pollution,

- Antibiotic usage and resistance,

- CO_2 and methane emissions, all of which together yield broken and ecosystem-degrading food-production systems.

Carbon Farming

Land management is the second largest contributor to carbon dioxide emissions on planet earth. Agriculture is the ONE sector that has the ability to transform from a net emitter of CO_2 to a net sequesterer of CO_2 — there is no other human managed realm with this potential. Common agricultural practices, including driving a tractor, tilling the soil, over-grazing, using fossil fuel based fertilizers, pesticides and herbicides result in significant carbon dioxide release. Alternatively, carbon can be stored long term (decades to centuries or more) beneficially in soils in a process called soil carbon sequestration. Carbon Farming involves implementing practices that are known to improve the rate at which CO_2 is removed from the atmosphere and converted to plant material and/or soil organic matter.

Carbon farming is successful when carbon gains resulting from enhanced land management and/or conservation practices exceed carbon losses.

Carbon Farming Practices

Recent studies demonstrate the efficacy of several carbon-beneficial agricultural practices in increasing soil carbon sequestration. Compost use has been shown to increase the amount of carbon stored in both grassland and cropland soils and has important co-benefits, such as increased primary productivity and water-holding capacity. Restoration of riparian areas on working lands has the capacity to sequester significant amounts of carbon. There are at least thirty-two on-farm Natural Resource Conservation Service (NRCS) conservation practices that are known to improve soil health and sequester carbon, while producing important co-benefits: increased water retention, hydrological function, biodiversity, and resilience.

Carbon Farm Planning

We start with the creation a Carbon Farm Plan (CFP), where our team works with a farmer or rancher to assess all the opportunities for GHG reduction and carbon sequestration on their property. A set of online tools (COMET) developed by researchers at Colorado State University, NRCS, CCI and the Marin Carbon Project, allows the quantification of GHG benefits. When we implement carbon farming, we also address many of ecosystem health impacts related to agriculture, including: groundwater and surface water degradation. Converting manure and other organic waste into high-quality compost, avoids the methane and air quality issues of conventional on-farm nutrient and waste management, and, improving soil health and soil organic matter directly improves the water holding capacity of soils, as we have seen first-hand on our demonstration farms across California.

Carbon Farming Implementation

The Carbon Cycle Institute has developed a model framework for land management that emphasizes carbon as the organizing principle. Land management within this framework leads to enhanced rates of carbon capture, increases the provision of important ecosystem services (especially water), and mitigates climate change. The framework relies on sound policies, public-private partnerships, quantification methodologies and innovative financing mechanisms that ultimately empower local organizations to efficiently implement on-the-ground, science-based solutions. Resource Conservation

Districts (RCDs) are an essential component of this framework. RCDs act as hubs that foster local partnerships to develop and implement carbon farming plans and practices in their districts, Several RCDs across California are building local partnerships, creating Carbon Farm Plans and engaging producers in carbon farming. it is critical to strengthen the capacity of RCDs and local agricultural support organizations to scale carbon farming to achieve measurable carbon capture, and address climate change and agricultural resilience, through both mitigation and adaptation.

Implications for our Climate

According to Marin Carbon Project research, sequestration of just one metric ton per hectare on half the rangeland area in California would offset 42 million metric tons of CO2e, an amount equivalent to the annual greenhouse gas emissions from energy use for all commercial and residential sectors in California.

The Carbon Cycle

Carbon constantly cycles through five pools on planet earth. Light energy coming from our sun functions as the fuel for the carbon cycle. The carbon cycle is a critical natural process that moves carbon through our atmosphere, biosphere, pedosphere, lithosphere, and oceans.

Human activity has tipped the balance of the carbon cycle through extracting enormous quantities of deeply sequestered fossil carbon as fossil fuels. These dense forms of carbon, when burned, release massive amounts of energy and carbon dioxide.

More carbon dioxide is now being released than the earth's land-based plant life and oceans can naturally reabsorb. The excess carbon dioxide has formed a blanket in our atmosphere—trapping the sun's heat and changing our climate, as seen in shifts in our earth's jet stream, ocean currents, and air temperature. Rainfall patterns are changing and glaciers (water storage for many communities) are melting quickly.

We have an opportunity to restore balance within the carbon cycle in a way that will ameliorate climate change, build resilience to drought and increase our agricultural productivity naturally. This natural solution is called Carbon Farming.

Agricultural Sustainability

Concerns about sustainability in agricultural systems centre on the need to develop technologies and practices that do not have adverse effects on environmental goods and services, are accessible to and effective for farmers, and lead to improvements in food productivity. Despite great progress in agricultural productivity in the past half-century, with crop and livestock productivity strongly driven by increased use of fertilizers, irrigation water, agricultural machinery, pesticides and land, it would be over-optimistic to assume that these relationships will remain linear in the future. New approaches are needed that will integrate biological and ecological processes into food production, minimize the use of those non-renewable inputs that cause harm to the environment or to the health of farmers and consumers, make productive use of the knowledge and skills of farmers, so substituting human capital for costly external inputs, and make productive use of people's collective capacities to work together to solve common agricultural and natural resource problems, such as for pest, watershed, irrigation, forest and credit management. These principles help to build important capital assets for agricultural systems: natural; social; human; physical; and financial capital. Improving natural capital is a central aim, and dividends can come from making the best use of the genotypes of crops and animals and the ecological conditions under which they are grown or raised. Agricultural sustainability suggests a focus on both genotype improvements through the full range of modern biological approaches and improved understanding of the benefits of ecological and agronomic management, manipulation and redesign. The ecological management of agro ecosystems that addresses energy flows, nutrient cycling, population-regulating mechanisms and system resilience can lead to the redesign of agriculture at a landscape scale. Sustainable agriculture outcomes can be positive for food productivity, reduced pesticide use and carbon balances. Significant challenges, however, remain to develop national and international policies to support the wider emergence of more sustainable forms of agricultural production across both industrialized and developing countries.

The Context for Agricultural Sustainability

The interest in the sustainability of agricultural and food systems can be traced to environmental concerns that began to appear in the 1950s–1960s. However, ideas about sustainability date back at least to the oldest surviving writings from China, Greece and Rome. Today, concerns about sustainability centre on the need to develop agricultural technologies and practices that:

- Do not have adverse effects on the environment (partly because the environment is an important asset for farming),

- Are accessible to and effective for farmers, and

- Lead to both improvements in food productivity and have positive side effects on environmental goods and services. Sustainability in agricultural systems incorporates concepts of both resilience (the capacity of systems to buffer shocks and stresses) and persistence (the capacity of systems to continue over long periods), and addresses many wider economic, social and environmental outcomes.

In recent decades, there has been remarkable growth in agricultural production, with increases in food production across the world since the beginning of the 1960s. Since then, aggregate world food production has grown by 145%. In Africa it rose by 140%, in Latin America by almost 200% and in Asia by 280%. The greatest increases have been in China, where a fivefold increase occurred, mostly during the 1980s–1990s. In industrialized countries, production started from a higher base; yet it still doubled in the USA over 40 years and grew by 68% in Western Europe.

Over the same period, world population has grown from three billion to more than six billion, imposing an increasing impact of the human footprint on the Earth as consumption patterns change. Again though, per capita agricultural production has outpaced population growth: for each person today, there is an additional 25% more food compared with in 1960. These aggregate figures, however, hide important regional differences. In Asia and Latin America, per capita food production increased by 76 and 28%, respectively. Africa, though, has fared badly, with food production per person 10% lower today than in 1960. China, again, performs best, with a trebling of per capita food production over the same period. These agricultural production gains have lifted millions out of poverty and provided a platform for rural and urban economic growth in many parts of the world.

However, these advances in aggregate productivity have not brought reductions in the incidence of hunger for all. In the early twenty-first century, there are still more than 800 million people hungry and lacking adequate access to food. A third is in East and Southeast Asia, another third in South Asia, a quarter in sub-Saharan Africa and 5% each in Latin America/Caribbean and in North Africa/Near East. Nonetheless, there has been progress, as incidence of undernourishment was 960 million in 1970, comprising a third of all people in developing countries at the time.

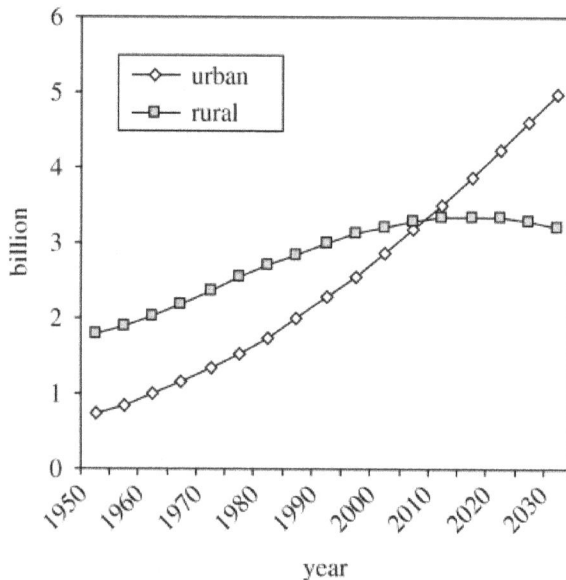

year

Rural and urban world population.

Despite this progress in food output, it is probable that food-related ill health will remain widespread for many people. As world population continues to increase, until at least the mid-twenty-first century, the absolute demand for food will also increase. Increasing incomes will also mean that people will have more purchasing power and this will increase the demand for food. But as diets change, demand for the types of food will also shift radically, with large numbers of people going through the nutrition transition. In particular, increasing urbanization means people are more likely to adopt new diets, particularly consuming more meat, fats and refined cereals, and fewer traditional cereals, vegetables and fruits.

As a result of these transitions towards calorie-rich diets, obesity, hypertension and type II diabetes have emerged as serious threats to health in most industrialized countries. A total of 20–25% of adults across Europe and North America are now classed as clinically obese (body mass index greater than 30kgm−2). In some developing countries, including Brazil, Colombia, Costa Rica, Cuba, Chile, Ghana, Mexico, Peru and Tunisia, overweight people now outnumber the hungry. Diet-related illness now has severe and costly public health consequences. According to the comprehensive Eurodiet study, 'disabilities associated with high intakes of saturated fat and inadequate intakes of vegetable and fruit, together with a sedentary lifestyle, exceed the cost of tobacco use'. Some problems arise from nutritional deficiencies of iron, iodide, folic acid, vitamin D and omega-3 polyunsaturated fatty acids, but most are due to excess consumption of energy and fat (causing obesity), sodium as salt (high blood pressure), saturated and trans fats (heart disease) and refined sugars.

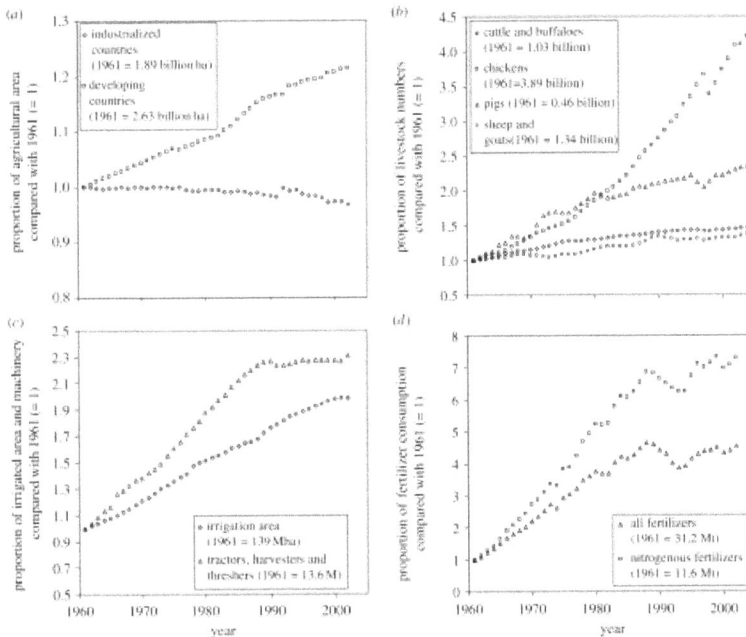

(a) Agricultural area (1961–2002). (b) Head of livestock, world (1961–2004). (c) Irrigated area and agricultural machinery, world (1961–2002). (d) World fertilizer consumption (1961–2002).

An important change in the world food system will come from the increased consumption of livestock products. Meat demand is expected to rise rapidly with economic growth and this will change many farming systems. Livestock are important in mixed production systems, using foods and by-products that would not have been consumed by humans. But increasingly animals are raised intensively and fed with cheap and energetically inefficient cereals and oils. In industrialized countries, 73% of cereals are fed to animals; in developing countries, some 37% are used in this way. Currently, per capita annual demand in industrialized countries is 550kg of cereal and 78kg of meat. By contrast, in developing countries, it is only 260kg of cereal and 30kg of meat.

At the same time as these recent changes in agricultural productivity, consumer behaviour over food and the political economy of farming and food, agricultural systems are now recognized to be a significant source of environmental harm. Since the early 1960s, the total agricultural area has expanded by 11% from 4.5 to 5 billion ha and arable area from 1.27 to 1.4 billion ha. In industrialized countries, agricultural area has fallen by 3%, but has risen by 21% in developing countries. Livestock production has also increased with a worldwide fourfold increase in numbers of chickens, twofold increase in pigs and 40–50% increase in numbers of cattle, sheep and goats.

During this period, the intensity of production on agricultural lands has also risen substantially. The area under irrigation and number of agricultural machines has grown by approximately twofold and the consumption of all fertilizers by fourfold (nitrogen fertilizers by sevenfold). The use of pesticides in agriculture has also increased dramatically and now amounts to some 2.56 billion kgyr−1. In the early twenty-first century, the annual value of the global market was US$25 billion, of which some US$3 billion of sales was in developing countries. Herbicides account for 49% of use, insecticides 25%, fungicides 22% and others approximately 3%. A third of the world market by value is in the USA, which represents 22% of active ingredient use. In the USA, though, large amounts of pesticide are used in the home/garden (17% by value) and in industrial, commercial and government settings (13% by value).

Table: World and US use of pesticide active ingredients (mean for 1998–1999).

Pesticide use	World pesticide use (million kga.i.)	%	US pesticide use (million kga.i.)	%
Herbicides	948	37	246	44
Insecticides	643	25	52	9
Fungicides	251	10	37	7
Other	721	28	219	40
Total	2563	100	554	100

These factors of production have had a direct impact on world food production. There are clear and significant relationships between fertilizer consumption, number of agricultural machines, irrigated area, agricultural land area and arable area with total world food production (comprising all cereals, coarse grains, pulses, roots and tubers, and oil crops). The inefficient use of some of these inputs has, however, led to considerable environmental harm. Increased agricultural area contributes substantially to the loss of habitats, associated biodiversity and their valuable environmental services. Approximately 30–80% of nitrogen applied to farmland escapes to contaminate water systems and the atmosphere as well as increasing the incidence of some disease vectors. Irrigation water is often used inefficiently and causes waterlogging and salinization, as well as diverts water from other domestic and industrial users; and agricultural machinery has increased the consumption of fossil fuels in food production.

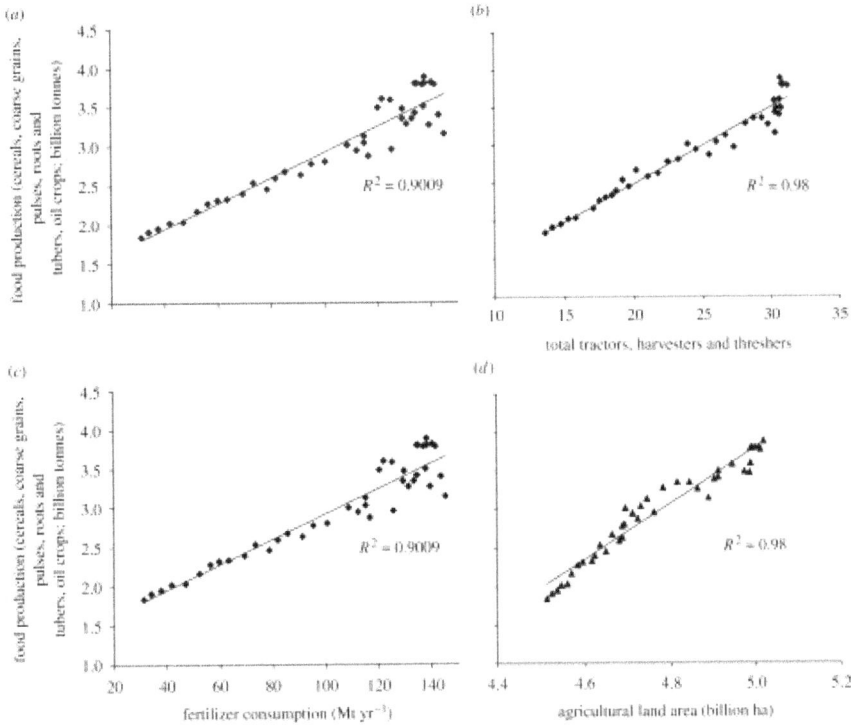

In figure, (a) Relationship between all fertilizers applied and world plant food production. (b) Relationship between world agricultural machinery and world plant food production. (c) Relationship between world irrigation area and world plant food production. (d) Relationship between world agricultural land area and world plant food production.

Figure clearly shows the past effectiveness of these factors of production in increasing agricultural productivity. One argument is to suggest that the persistent world food crisis indicates a need for substantially greater use of these inputs. But it would be both simplistic and optimistic to assume that all these relationships will remain linear in the future and that gains will continue at the previous rates. This would assume a continuing supply of these factors and inputs, and that the environmental costs of their use will be small. There is also growing evidence to suggest that this approach to agricultural growth has reached critical environmental limits, and that the aggregate costs in terms of lost or foregone benefits from environmental services are too great for the world to bear. The costs of these environmental problems are often called externalities as they do not appear in any formal accounting systems. Yet many agricultural systems themselves are now suffering because key natural assets that they require to be plentiful are being undermined or diminished.

Agricultural systems in all parts of the world will have to make improvements. In many, the challenge is to increase food production to solve immediate problems of hunger. In others, the focus will be more on adjustments that maintain food production while increasing the flow of environmental goods and services. World population is set to continue to increase for approximately another 40 years to approximately 2040–2050, and then is likely to stabilize or fall owing to changes in fertility patterns. The high-fertility projection by the UN is unlikely to arise, as shifts towards lower fertility are already occurring in many countries worldwide and so there are very real prospects of world population eventually falling over one to two centuries after the maximum is reached. This suggests that the agricultural and food challenge is likely to be more acute in the next half-century, and thereafter qualitatively change according to people's aggregate consumption patterns.

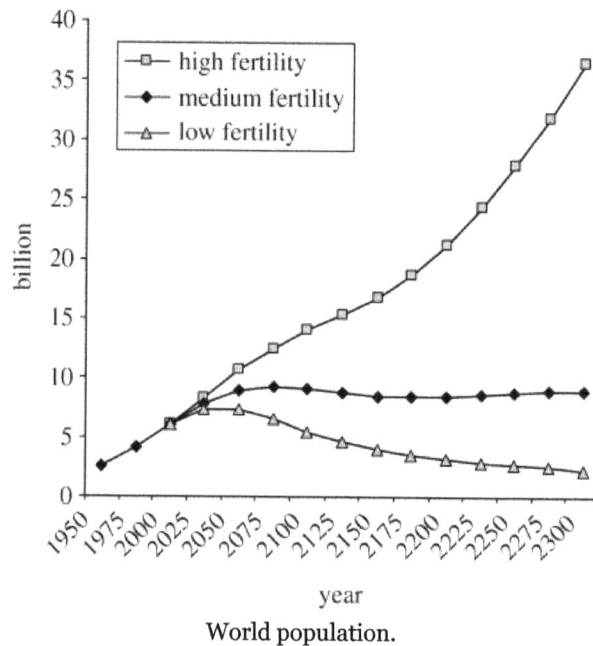

World population.

Sustainable Agriculture

What, then, do we now understand by agricultural sustainability? Many different expressions have come to be used to imply greater sustainability in some agricultural systems over prevailing ones (both pre-industrial and industrialized). These include biodynamic, community based, ecoagriculture, ecological, environmentally sensitive, extensive, farm fresh, free range, low input, organic, permaculture, sustainable and wise use. There is continuing and intense debate about whether agricultural systems using some of these terms can qualify as sustainable.

Systems high in sustainability can be taken as those that aim to make the best use of environmental goods and services while not damaging these assets. The key principles for sustainability are to:

i)   Integrate biological and ecological processes such as nutrient cycling, nitrogen fixation, soil regeneration, allelopathy, competition, predation and parasitism into food production processes,

ii) Minimize the use of those non-renewable inputs that cause harm to the environment or to the health of farmers and consumers,

iii) Make productive use of the knowledge and skills of farmers, thus improving their self-reliance and substituting human capital for costly external inputs, and

iv) Make productive use of people's collective capacities to work together to solve common agricultural and natural resource problems, such as for pest, watershed, irrigation, forest and credit management.

The idea of agricultural sustainability, though, does not mean ruling out any technologies or practices on ideological grounds. If a technology works to improve productivity for farmers and does not cause undue harm to the environment, then it is likely to have some sustainability benefits. Agricultural systems emphasizing these principles also tend to be multifunctional within landscapes and economies. They jointly produce food and other goods for farmers and markets, but also contribute to a range of valued public goods, such as clean water, wildlife and habitats, carbon sequestration, flood protection, groundwater recharge, landscape amenity value and leisure/tourism. In this way, sustainability can be seen as both relative and case dependent and implies a balance between a range of agricultural and environmental goods and services.

As a more sustainable agriculture seeks to make the best use of nature's goods and services, technologies and practices must be locally adapted and fitted to place. These are most likely to emerge from new configurations of social capital, comprising relations of trust embodied in new social organizations, new horizontal and vertical partnerships between institutions, and human capital comprising leadership, ingenuity, management skills and capacity to innovate. Agricultural systems with high levels of social and human assets are more able to innovate in the face of uncertainty. This suggests that there likely to be many pathways towards agricultural sustainability, and further implies that no single configuration of technologies, inputs and ecological management is more likely to be widely applicable than the other. Agricultural sustainability implies the need to fit these factors to the specific circumstances of different agricultural systems.

A common, though erroneous, assumption about agricultural sustainability is that it implies a net reduction in input use, thus making such systems essentially extensive (they require more land to produce the same amount of food). Recent empirical evidence shows that successful agricultural sustainability initiatives and projects arise from shifts in the factors of agricultural production (e.g. from use of fertilizers to nitrogen-fixing legumes; from pesticides to emphasis on natural enemies; from ploughing to zero-tillage). A better concept than extensive is one that centres on intensification of resources, making better use of existing resources (e.g. land, water, biodiversity) and technologies. The critical question centres on the 'type of intensification'. Intensification using natural, social and human capital assets, combined with the use of best available technologies and inputs (best genotypes and best ecological management) that minimize or eliminate harm to the environment, can be termed 'sustainable intensification'.

Effects of Sustainable Agriculture on Yields

One persistent question regarding the potential benefits of more sustainable agro ecosystems centres on productivity trade-offs. If environmental goods and services are to be protected or improved, what then happens to productivity? If it falls, then more land will be required to produce the same amount of food, thus resulting in further losses of natural capital. As indicated earlier, the challenge is to seek sustainable intensification of all resources in order to improve food production. In industrialized farming systems, this has proven impossible to do with organic production systems, as food productivity is lower for both crop and livestock systems. Nonetheless, there are now some 3Mha of agricultural land in Europe managed with certified organic practices. Some have led to lower energy use (though lower yields too), others to better nutrient retention and some greater nutrient losses.

Many other farmers have adopted integrated farming practices, which represent a step or several steps towards sustainability. What has become increasingly clear is that many modern farming systems are wasteful, as integrated farmers have found they can cut down many purchased inputs without losing out on profitability. Some of these cuts in use are substantial, others are relatively small. By adopting better targeting and precision methods, there is less wastage and more benefit to the environment. They can then make greater cuts in input use once they substitute some regenerative technologies for external inputs, such as legumes for inorganic fertilizers or predators for pesticides. Finally, they can replace some or all external inputs entirely over time once they have learned their way into a new type of farming characterized by new goals and technologies.

However, it is in developing countries that some of the most significant progress towards sustainable agro ecosystems has been made in the past decade. The largest study comprised the analysis of 286 projects in 57 countries. This involved the use of both questionnaires and published reports by projects to assess changes over time. As in earlier research, data were triangulated from several sources and cross-checked by external reviewers and regional experts. The study involved analysis of projects sampled once in time (n=218) and those sampled twice over a 4-year period (n=68). Not all proposed cases were accepted for the dataset and rejections were based on a strict set of criteria. As this was a purposive sample of 'best practice' initiatives, the findings are not representative of all developing country farms.

Table contains a summary of the location and extent of the 286 agricultural sustainability projects across the eight categories of FAO farming systems in the 57 countries. In all, some 12.6 million farmers on 37Mha were engaged in transitions towards agricultural sustainability in these 286 projects. This is just over 3% of the total cultivated area (1.136Mha) in developing countries. The largest number of farmers was in wetland rice-based systems, mainly in Asia (category 2), and the largest area was in dualistic mixed systems, mainly in southern Latin America (category 6). This study showed that agricultural sustainability was spreading to more farmers and hectares. In the 68 randomly re-sampled projects from the original study, there was a 54% increase over the 4 years in the number of farmers and 45% in the number of hectares. These resurveyed projects comprised 60% of the farmers and 44% of the hectares in the original sample of 208 projects.

Table: Summary of adoption and impact of agricultural sustainability technologies and practices on 286 projects in 57 countries.

FAO farm system category	No. of farmers adopting	No. of hectares under sustainable agriculture	Average % increase in crop yields
Smallholder irrigated	177287	357940	129.8 (±21.5)
Wetland rice	8711236	7007564	22.3 (±2.8)
Smallholder rainfed humid	1704958	1081071	102.2 (±9.0)
Smallholder rainfed highland	401699	725535	107.3 (±14.7)
Smallholder rainfed dry/cold	604804	737896	99.2 (±12.5)
Dualistic mixed	537311	26846750	76.5 (±12.6)
Coastal artisanal	220000	160000	62.0 (±20.0)
Urban-based and kitchen garden	207479	36147	146.0 (±32.9)
All projects	12564774	36952903	79.2 (±4.5)

For the 360 reliable yield comparisons from 198 projects, the mean relative yield increase was 79% across the very wide variety of systems and crop types. However, there was a widespread in results. While 25% of projects reported relative yields greater than 2.0 (i.e. 100% increase), half of all the projects had yield increases between 18 and 100%. The geometric mean is a better indicator of the average for such data with a positive skew, but this still shows a 64% increase in yield. However, the average hides large and statistically significant differences between the main crops. In nearly all cases, there was an increase in yield with the project. Only in rice there were three reports where yields decreased, and the increase in rice was the lowest (mean=1.35), although it constituted a third of all the crop data. Cotton showed a similarly small mean yield increase.

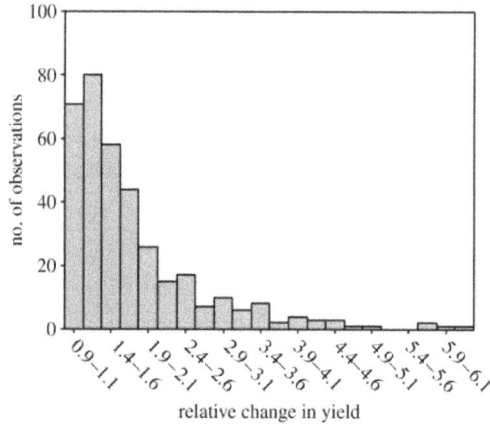

Histogram of change in crop yield after or with project, compared with before or without project (n=360, mean =1.79, s.d.=0.91, median=1.50, geometric mean=1.64).

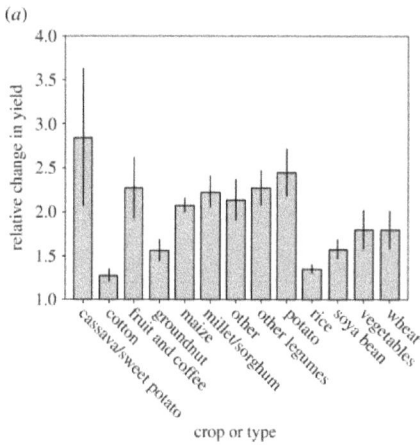

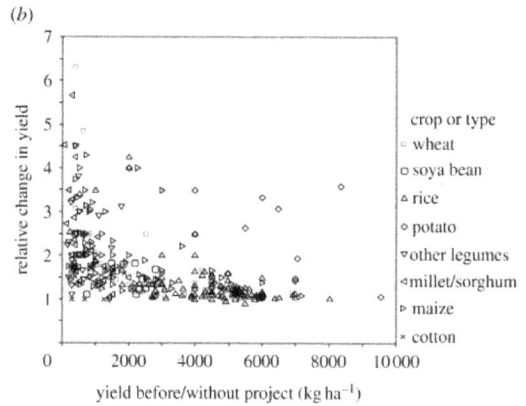

(a) Mean changes in crop yield after or with project, compared with before or without project. Vertical lines indicate ±s.e.m. 'Other' group consists of sugar cane (n=2), quinoa (1), oats (2). (b) Relationship between relative changes in crop yield after (or with project) to yield before (or without project). Only field crops with n>9 shown.

These sustainable agroecosystems also have positive side effects, helping to build natural capital, strengthen communities (social capital) and develop human capacities. Examples of positive side effects recently recorded in various developing countries include:

- improvements to natural capital, including increased water retention in soils, improvements in water table (with more drinking water in the dry season), reduced soil erosion combined with improved organic matter in soils, leading to better carbon sequestration, and increased agro biodiversity.

- improvements to social capital, including more and stronger social organizations at local level, new rules and norms for managing collective natural resources, and better connectedness to external policy institutions.

- improvements to human capital, including more local capacity to experiment and solve own problems, reduced incidence of malaria in rice-fish zones, increased self-esteem in formerly marginalized groups, increased status of women, better child health and nutrition, especially in dry seasons, and reversed migration and more local employment.

What we do not know, however, is the full economic benefits of these spin-offs. In many industrialized countries, agriculture is now assumed to contribute very little to gross domestic product, leading many commentators to assume that agriculture is not important for modernized economies. But such a conclusion is a function of the fact that very few measures are being made of the positive side effects of agriculture. In poor countries, where financial support is limited and markets weak, then people rely even more on the value they can derive from the natural environment and from working together to achieve collective outcomes.

Effects of Sustainable Agriculture on Pesticide use and Yields

Recent IPM programmes, particularly in developing countries, are beginning to show how pesticide use can be reduced and pest management practices can be modified without yield penalties. In principle, there are four possible trajectories of impact if IPM is introduced:

- Pesticide use and yields increase (A),

- Pesticide use increases, but yields decline (B),

- Both pesticide use and yields fall (C), and

- Pesticide use declines, but yields increase (D).

The assumption in modern agriculture is that pesticide use and yields are positively correlated. For IPM, the trajectory moving into sector A is therefore unlikely but not impossible, for example in low-input systems. What is expected is a move into sector C. While a change into sector B would be against economic rationale, farmers are unlikely to adopt IPM if their profits would be lowered. A shift into sector D would indicate that current pesticide use has negative yield effects or that the amount saved from pesticides is reallocated to other yield-increasing inputs. This could be possible with excessive use of herbicides or when pesticides cause outbreaks of secondary pests, such as observed with the brown plant hopper in rice.

Figure shows data from 62 IPM initiatives in 26 developing and industrialized countries. The 62 IPM initiatives have some 5.4 million farm households on 25.3Mha. The evidence on pesticide use is derived from data on both the number of sprays per hectare and the amount of active ingredient used per hectare. This analysis does not include recent evidence on the effect of some genetically modified crops, some of which result in reductions in the use of herbicides and pesticides, and some of which have led to increases.

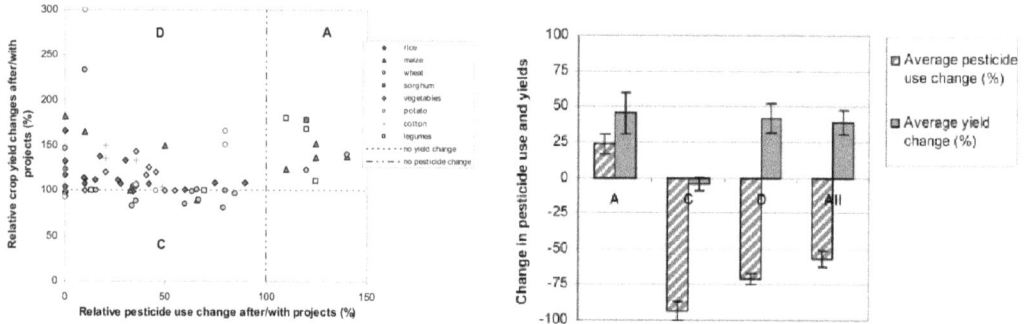

(a) Association between pesticide use and crop yields (data from 80 crop combinations, 62 projects, 26 countries). (b) Changes in pesticide use and yields in 62 projects (A: n=10; C: n=5; D: n=47).

There is only one sector B case reported in recent literature. Such a case has recently been reported from Java for rice farmers. The cases in sector C, where yields fall slightly while pesticide use falls dramatically, are mainly cereal-farming systems in Europe, where yields typically fall to some 80% of current levels while pesticide use is reduced to 10–90% of current levels. Sector A contains 10 projects where total pesticide use has indeed increased in the course of IPM introduction. These are mainly in zero-tillage and conservation agriculture systems, where reduced tillage creates substantial benefits for soil health and reduced off-site pollution and flooding costs. These systems usually require increased use of herbicides for weed control, though there are some examples of organic zero-tillage systems. Over 60% of the projects are in category D where pesticide use declines and yields increase. While pesticide reduction is to be expected, as farmers substitute pesticides by information, yield increase induced by IPM is a more complex issue. It is probable, for example, that farmers who receive good quality field training will not only improve their pest management skills but also become more efficient in other agronomic practices such as water, soil and nutrient management. They can also invest some of the cash saved from pesticides in other inputs such as higher quality seeds and inorganic fertilizers.

Effects on Carbon Balances

The 1997 Kyoto Protocol to the UN Framework Convention on Climate Change established an international policy context for the reduction of carbon emissions and increases in carbon sinks in order to address the global challenge of anthropogenic interference with the climate system. It is clear that both emission reductions and sink growth will be necessary for mitigation of current climate change trends. A source is any process or activity that releases a greenhouse gas, or aerosol or a precursor of a greenhouse gas into the atmosphere, whereas a sink is such mechanism that removes these from the atmosphere. Carbon sequestration is defined as the capture and secure storage of carbon that would otherwise be emitted to or remain in the atmosphere. Agricultural systems emit carbon through the direct use of fossil fuels in food production, the indirect use of embodied energy in inputs that are energy intensive to manufacture, and the cultivation of soils and/or soil erosion resulting in the loss of soil organic matter. Agriculture also contributes to climate change through the emissions of methane from irrigated rice systems and ruminant livestock. The direct effects of land use and land-use change (including forest loss) have led to a net emission of 1.7GtCyr^{-1} in the 1980s and 1.6GtCyr^{-1} in the 1990s.

On the other hand, agriculture is also an accumulator of carbon when organic matter is accumulated in the soil, and when above-ground biomass acts either as a permanent sink or is used as an energy source that substitutes for fossil fuels and thus avoids carbon emissions. There are 3 main mechanisms and 21 technical options by which positive actions can be taken by farmers by:

i.   Increasing carbon sinks in soil organic matter and above-ground biomass,

ii. Avoiding carbon dioxide or other greenhouse gas emissions from farms by reducing direct and indirect energy use, and

iii. Increasing renewable energy production from biomass that either substitutes for consumption of fossil fuels or replacing inefficient burning of fuel wood or crop residues, and so avoids carbon emissions.

Table: Mechanisms for increasing carbon sinks and reducing CO_2 and other greenhouse gas emissions in agricultural systems.

Mechanism A	Increase carbon sinks in soil organic matter and above-ground biomass replace inversion ploughing with conservation- and zero-tillage systems adopt mixed rotations with cover crops and green manures to increase biomass additions to soil adopt agroforestry in cropping systems to increase above-ground standing biomass minimize summer fallows and periods with no ground cover to maintain soil organic matter stocks use soil conservation measures to avoid soil erosion and loss of soil organic matter apply composts and manures to increase soil organic matter stocks improve pasture/rangelands through grazing, vegetation and fire management both to reduce degradation and increase soil organic matter cultivate perennial grasses (60–80% of biomass below ground) rather than annuals (20% below ground) restore and protect agricultural wetlands convert marginal agricultural land to woodlands to increase standing biomass of carbon.
Mechanism B	Reduce direct and indirect energy use to avoid greenhouse gas emissions (CO_2, CH_4 and N_2O) conserve fuel and reduce machinery use to avoid fossil fuel consumption use conservation- or zero-tillage to reduce CO_2 emissions from soils adopt grass-based grazing systems to reduce methane emissions from ruminant livestock use composting to reduce manure methane emission substitute biofuel for fossil fuel consumption reduce the use of inorganic N fertilizers (as manufacturing is highly energy intensive), and adopt targeted- and slow-release fertilizers use IPM to reduce pesticide use (avoid indirect energy consumption).

Mechanism C	Increase biomass-based renewable energy production to avoid carbon emissions cultivate annual crops for biofuel production such as ethanol from maize and sugar cane cultivate annual and perennial crops, such as grasses and coppiced trees, for combustion and electricity generation, with crops replanted each cycle for continued energy production use biogas digesters to produce methane, so substituting for fossil fuel sources use improved cookstoves to increase efficiency of biomass fuels.

The potential annual contributions being made in the 286 projects to carbon sink increases in soils and trees were calculated, using an established methodology. As the focus is on what sustainable methods can do to increase quantities of soil and above-ground carbon, no account was taken of existing stocks of carbon. Soil carbon sequestration is corrected for climate, as rates are higher in humid when compared with dry zones and generally higher in temperate than tropical areas.

Table: Summary of potential carbon sequestered in soils and above-ground biomass in the 286 projects. (±s.e. in brackets.)

FAO farm system category	Carbon sequestered per hectare (tCha^{-1}yr^{-1})	Total carbon sequestered (MtCyr^{-1})	Carbon sequestered per household (tCyr^{-1})
Smallholder irrigated	0.15 (±0.012)	0.011	0.06
Wetland rice	0.34 (±0.035)	2.53	0.29
Smallholder rainfed humid	0.46 (±0.034)	0.34	0.20
Smallholder rainfed highland	0.36 (±0.022)	0.23	0.56
Smallholder rainfed dry/cold	0.26 (±0.035)	0.20	0.32
Dualistic mixed	0.32 (±0.023)	8.03	14.95
Coastal artisanal	0.20 (±0.001)	0.032	0.15
Urban-based and kitchen garden	0.24 (±0.061)	0.015	0.07
Total	0.35 (±0.016)	11.38	0.91

These projects were potentially sequestering 11.4MtCyr^{-1} on 37Mha. The average gain was 0.35tCha^{-1}yr^{-1}, with an average per household gain of 0.91tCyr^{-1}. The per hectare gains vary from 0.15tCha^{-1}yr^{-1}for smallholder irrigated systems (category 1) to 0.46tCha^{-1}yr^{-1} for category three systems. For most systems, per households gains were in the range 0.05–0.5tCyr^{-1}, with the much larger farms of southern Latin America using zero-tillage and conservation agriculture achieving the most at 14.9tCyr^{-1}. Such gains in carbon may offer new opportunities for income generation under carbon trading schemes.

Permaculture

Permaculture can be understood as the growth of agricultural ecosystems in a self-sufficient and sustainable way.

This form of agriculture draws inspiration from nature to develop synergetic farming systems based on crop diversity, resilience, natural productivity, and sustainability. Still, since the early 1980s, the pre-conceived idea of permaculture extended to a systemic approach that goes far beyond the agricultural domain.

Nowadays synonymous with permanent culture in its broadest sense, permaculture is a global ethic method for designing integrated systems based on the idea of sustainable development. Therefore, human activities must consider natural ecosystems and operate in harmony with them.

Permacultural Design: The Principles and Techniques of Permaculture

Based on the precise observation of how ecosystems work (particularly in terms of productivity and efficiency), permaculture draws on non-fixed modes of design that are adaptable to the fields of application. The result is a method of universal principles known as "permacultural design".

The permacultural design (the word *design* here includes the notions of project and process of realization) is set up from three founding ethical requirements:

- Preservation of the environment and biodiversity;
- Willingness to build a community for individual and collective well-being;
- Sharing of resources and equitable redistribution of excesses (for the benefit of humans and the environment).

The method itself is based on:

- An overall understanding of issues and systems;
- The analysis of the connection modes between the elements of a system;
- The application, to deficient systems, of solutions derived from operational and proven systems;
- The analysis of natural ecosystems to correct errors in the implantation of human activity and planning for optimal integration;
- The inclusion of people new to permaculture in the process.

Permacultural design implements many solutions inspired by scientific ecology, biomimicry, and empirical practices developed over the ages by traditional societies.

Examples and Practices of Permaculture

- Agricultural example: no-till practices that ensure the preservation of soil balance and long-term fertility.

- Habitats example: buildings made of biodegradable local materials that are energy-efficient and have a minimal ecological footprint (individual houses made of straw and clay in the Netherlands).

- Economy example: the development of community organizations such as SEL (Local Exchange System), generating short circuits, social links, solidarity, and community cohesion.

Permaculture and Agroecology: Differences and Similarities

Agroecology and permaculture are often confused, yet these two practices are different. Agro-ecology goes further than biological agricultural as it uses sustainable agricultural systems with techniques such as complementarity, composting or cultivation on mounds. Afterward, it integrates these systems in an ecological way by saving water, fighting against erosion etc.

The practices above can be found in permaculture too, but the latter is broader since it focuses not only on creating sustainable and resilient farming systems but also on integrating them into a broader life system where other variables are also accounted for.

Agroforestry

Agroforestry is the management and integration of trees, crops and/or livestock on the same plot of land and can be an integral component of productive agriculture. It may include existing native forests and forests established by landholders. It is a flexible concept, involving both small and large-sized land holdings.

Scientifically speaking, agroforestry is derived from ecology and is one of the three principal land-use sciences, the other two being agriculture and forestry. Agroforestry differs from the latter two principals by placing an emphasis on integration of and interactions among a combination of elements rather than just focussing on each element individually.

Agroforestry has a lot in common with intercropping (the practice of planting two or more crops on the same plot) with both practices placing an emphasis on interaction between different plant species. Generally speaking, both agrofrestry and intercropping can result in higher overall yields and reduced operational costs.

Trees for the Future

Benefits of Agroforestry

Over the past two decades, a number of studies have been carried out analysing the viability of agroforestry. The combined research has highlighted that agroforestry can reap substantial benefits both economically and environmentally, producing more output and proving to be more sustainable than forestry or agricultural monocultures. Agroforestry systems have already been adopted in many parts of the world.

Agroforestry systems can include the following benefits:

1. They can control runoff and soil erosion, thereby reducing losses of water, soil material, organic matter and nutrients.

2. They can maintain soil organic matter and biological activity at levels satisfactory for soil fertility. This depends on an adequate proportion of trees in the system- normally at least 20% crown cover of trees to maintain organic matter over systems as a whole.

3. They can maintain more favourable soil physical properties than agriculture, through organic matter maintenance and the effects of tree roots.

4. They can lead to more closed nutrient cycling than agriculture and hence to more efficient use of nutrients. This is true to an impressive degree for forest garden/farming systems.

5. They can check the development of soil toxicities, or reduce exiting toxicities-both soil acidification and salinization can be checked and trees can be employed in the reclamation of polluted soils.

6. They utilize solar energy more efficiently than monocultural systems different height plants, leaf shapes and alignments all contribute.

7. They can lead to reduced insect pests and associated diseases.

8. They can be employed to reclaim eroded and degraded land.

9. Agro forestry can augment soil water availability to land use systems. In dry regions, though, competition between trees and crops is a major problem.

10. Nitrogen-fixing trees and shrubs can substantially increase nitrogen inputs to agro forestry systems.

11. Trees can probably increase nutrient inputs to agro forestry systems by retrieval from lower soil horizons and weathering rock.

12. The decomposition of tree and pruning can substantially contribute to maintenance of soil fertility. The addition of high-quality tree prunings leads to large increase in crop yields.

13. The release of nutrients from the decomposition of tree residues can be synchronized with the requirements for nutrient uptake of associated crops. While different trees and crops will all have different requirement, and there will always be some imbalance, the addition of high quality prunings to the soil at the time of crop planting usually leads to a good degree of synchrony between nutrient release and demand.

14. In the maintenance of soil fertility under agro forestry, the role of roots is at least as important as that of above-ground biomass.

15. Agro forestry can provide a more diverse farm economy and stimulate the whole rural economy, leading to more stable farms and communities. Economics risks are reduced when systems produce multiple products.

As well as building on practices used in forestry and agriculture, agroforestry also works towards land protection and conservation through more effective protection of stock, control of soil erosion, salinity and water tables and a higher quality control of timber.

A denser, more-dependable tree covering can provide shelter to livestock during the warmer months allowing the animals can conserve energy. That same tree covering helps block out wind, helping to boost water retention levels that can help produce a more robust crop yield.

Agroforestry can improve land protection in the following areas:

1. Salinity and water table control: Salinity is mainly caused by rising water tables. Trees help to lower water tables, acting as pumps to take up water from the soil and then evaporating it to the atmosphere.

2. Soil erosion control: Soil erosion or loss results from the action of wind and water on unprotected soils. The forest canopy, roots and leaf litter all have a role in controlling soil erosion.

3. Water logging: Through water removal, established trees can substantially reduce water logging in their immediate area, which may result in improved land uses, e.g. pasture or crop.

Agroforestry can have immense benefits for the environment and the farmer (a detailed breakdown of agroforestry's main benefits can be found on AgriInfo's site). For farmers, the ability to maintain some sort of control over land and production in the face of climate change means agrofrestry could hold huge promise for the agricultural sector.

On an environmental level, agroforestry's ability to help prevent soil erosion while simultaneously aiding water retention and promoting soil fertility could help provide a solution for areas where rainfal is irregular or might become irregular due to climate change while dense plantations of trees would also help absorb CO_2 and regulate local temperature.

Agroecological Approaches

Agroecology as an Alternative Path to Industrialized Agriculture: Despite the impressive growing number of scientific work published in this field and the increasing global recognition the concept is enjoying, agroecology too often remains wrongly perceived as one particular set of agricultural practices which could substantially help increase agricultural sustainability but only in a few very specific, limited contexts, and therefore cannot pretend to be a credible solution at a global scale. Such a narrow view is far from reality. In terms of agricultural practices or farming systems, agroecology is rather a holistic approach consisting in realizing key principles through the context-specific design of strategies and techniques. But agroecology is not only an agricultural approach. It is also referred to as a science and a social movement. While agroecology first emerged as a science, trajectories between science, social movement and agricultural approach are very diverse depending on countries.

The concept of agroecology encompasses different meanings depending on the actors and the given socio-historical context, and is a living concept, submitted to permanent evolution. It is however possible to identify common features beyond that diversity. For example, the movement for agroecology builds on agroecological science and knowledge for promoting and practicing the agricultural approach. As argued by Wezel et al., agroecology is neither exclusively defined as scientific disciplines, nor exclusively as social movements or practices. It is a federative concept of actions, intermediate between the three dimensions.

Agroecology as a Science

As a first step, agroecology developed through an attempt to integrate the principles of ecology to the redefinition of agronomy. The term was first used in two scientific publications by Bensin, a Russian agronomist, for describing the use of ecological methods in research on commercial crop plants. In 1965, in what is probably the first book titled 'agroecology', the German ecologist/zoologist Tischler analyzed the different components such as plants, animals, soils, and climate, and their interactions within an agroecosystem as well as the impact of human agricultural management on these components, thus applying an approach combining ecology, especially the interactions among biological components at the field or agroecosystem level, and agronomy with a focus on the integration of agricultural management. Today's most frequent benchmark definition of agroecology as a science combining ecology and agronomy has been established by Altieri, entomologist from University of California Berkeley. He defines agroecology as "the application of ecological science to the study, design, and management of sustainable agriculture". Focused on the analysis of agroecosystems (communities of plants and animals interacting with their physical and chemical environments), agroecology hence aims at producing knowledge and practices which provide the means to make agriculture more sustainable. Put crudely, as a scientific discipline, agroecology can be understood as "the science behind sustainable agriculture", or the science of sustainable agriculture.

But while this definition remains widely used, since the 1930's the scope and nature of agroecology as a scientific discipline have broadened considerably, moving beyond the level of agroecosystems towards a larger focus on the whole food system (defined as a global network of food production, distribution and consumption), and developing a transdisciplinary approach, thus no more exclusively based on biotechnical sciences but also applying social sciences. This evolution can be well illustrated for example by Francis et al., who defined agroecology as "the integrative study of the ecology of the entire food systems, encompassing ecological, economic and social dimensions, or more simply the ecology of food systems". As a scientific discipline, agroecology is increasingly considered as the science of sustainable food systems.

Agroecology as a science is first and foremost based on the re-discovery and study of traditional peasant agricultures. This close relationship results from the recognition of the phenomenal sustainability that traditional peasant farming systems have demonstrated throughout the ages, and as a corollary of the treasures of knowledge they represent for achieving sustainability today and in the future, including in the context of climate change. The myriad of existing traditional systems indeed reveals a tremendous diversity of domesticated crop and animal species maintained and enhanced by soil, water and biodiversity management regimes nourished by complex traditional knowledge systems. These systems comprise a significant ingenious agricultural heritage reflecting the extreme diversity of agricultural systems adapted to different environments. They have not only fed much of the world population for centuries and continue to feed people in many parts of the planet, especially in developing countries, but undoubtedly also hold many of the potential answers to the production and natural resource conservation challenges affecting today's rural landscapes. Agroecology therefore strongly recognizes the crucial importance of preserving them.

The practice of agroecology as scientific discipline has allowed for the identification of key principles that form the foundation of agricultural sustainability. Literature on agroecology most often refers to the following five core principles:

- Increasing the recycling of biomass and achieving a balance in nutrients flow;

- Assuring favorable soil conditions, keeping the soil covered with mulch or cover crops, guaranteeing a high level of soil organic matter and an active soil biology;

- Minimizing nutrients losses from the system, through relatively closed rather than open system design;

- Promoting the functional biodiversity of the system, including within–and between-species diversity, above–and below-ground and landscape level biodiversity;

- Promoting increased biological interactions and synergisms among system components that can sponsor system services like regenerating soil fertility and providing pest management without resorting to external inputs.

Realizing these principles must notably lead to minimizing the use of non-renewable inputs that cause harm to the environment or to the health of farmers and consumers. The five 'historical' principles of agroecology have been theorized in a restrictive ecosystem perspective, intending to protect peasant agroecosystems from negative consequences of the Green Revolution and dependence on external inputs for promoting an endogenic dynamic of development, valorizing the use of local resources for supporting small-scale agriculture, more sustainable from a social and environmental point of view.

The historical principles mentioned above are widely accepted as core pillars of agro-ecology. However, identification of key principles remains a topic of debate and is subject to further theorization, especially when integrating broader social or political aspects of the agroecological paradigm. For example, based on criteria derived from the extensive literature on agroecology and sustainable agriculture, several authors including Altieri have highlighted a comprehensive list of 10 basic attributes that any agricultural system should exhibit in order to be considered sustainable. More recently, the Interdisciplinary Group of Research on 'Agroecology' of the Belgian Fonds de la Recherche Scientifique (Fnrs) (GIRAF) has proposed a conceptual framework completing the five historical principles by 8 additional ones, both based on the French National Institute for Agricultural Research (INRA) and its own work, thus proposing a total of 13 principles with the objective of guiding further work on agroecology. The conceptualization of these additional principles takes due account of the social ambitions of agroecology.

Agroecology as an Agricultural Approach

Since the 1970s, agroecology no longer referred simply to a scientific discipline or research area, but also to farming practices and a number of collective mobilizations (mainly in response to the Green Revolution). In terms of farming systems, agroecology could be synthetically defined as a holistic approach consisting in seeking to make agroecosystems economically, ecologically and socially more sustainable by realizing key agroecological principles (that are precisely understood as those which form the basis of agricultural sustainability) for meeting local needs. Agroecological farming indeed promotes community-oriented approaches that look after the subsistence needs of its members, and very much privilege the local: providing for local markets that shorten the circuits of food production and consumption, simultaneously avoiding the high energy needs of 'long-distance food'. It also seeks to increase resilience. Usually defined as the propensity of a system to retain its organizational structure and productivity following perturbation, resilience is a constant preoccupation of agroecology. Realizing agroecological principles consists primarily in mimicking natural processes, thus creating beneficial biological interactions and synergies among the components of the agroecosystem through multiple, context-specific combinations of strategies and practices that are designed, applied and managed primarily by farmers themselves, building first and foremost on their traditional knowledge and know-how.

Designing a Strategy for Managing a Transition

While agroecological principles have universal applicability, the technological forms through which they can be made operational depend on the prevailing environmental and socioeconomic conditions at each site. In other words, their concrete realization always requires contextspecific solutions, since they must adapt to local realities. As a process of transition towards more sustainable agricultural systems, agroecology consists therefore essentially in designing and applying an adequate strategy for managing the transition, one that can improve sustainability in the particular context considered, through means that are adapted to local conditions. As a starting point for designing such strategy, agroecology implies proceeding to a comprehensive diagnosis of sustainability challenges and conditions specific to the given context. Simply put, the question is: what are the priorities in this context for improving agricultural sustainability and how can they be concretely addressed.

This diagnosis requires a holistic approach. This means that all relevant aspects of sustainability, whether linked to food security, environmental protection and/or to community well-being, must be taken into account, recognizing the multifunctionality of agriculture. This also implies identifying all human (economic, social, cultural, political) and environmental constraints, as well as the ways through which those elements interact with each other, and mapping all assets (natural, social, human, physical and financial) locally available. Agricultural systems at all levels indeed rely on the value of services flowing from the total stock of assets that they influence and control. Moreover, the holistic approach means defining expected benefits in the short, medium and long term and going beyond the level of the plot or the farming system, since many sustainability challenges also depend on upper spatial scales. This is the case for example of environmental challenges such as the sustainability of varietal resistance within territories, biodiversity maintenance at the landscape level, GHG emissions at the global level, etc. The need for analysis at the landscape or territory level implies thinking in terms of collective actions, thus requiring coordination between different actors. Coordination among actors is particularly important in case of conflicting expectations as to the use of land, water or other natural resources. This explains why ensuring responsible governance of natural resources is important from an agroecological perspective. Indeed, sustainable management of these resources necessarily implies (re)conciliating in a sustainable way the actors' different expectations and interests as to the use of resources. All these different elements are essential components of the agroecological equation consisting in designing the best options for improving sustainability.

Solving this equation requires conceiving farming systems that rely primarily on functionalities given by ecosystems and built on traditional local knowledge. The relevance of making the best use of traditional knowledge for designing agroecological systems is obvious since this knowledge is intrinsically adapted to local conditions in a given environment. And it crystallizes an extreme diversity of options that for centuries have helped farmers to sustainably manage harsh environments and to meet their subsistence needs, without depending on mechanization, chemical fertilizers, pesticides, or other technologies of modern agricultural science. However, agroecology does not imply excluding all modern technologies on ideological grounds. If a technology works to improve productivity for farmers and does not cause undue harm to the environment, then it is likely to have some sustainability benefits. Agroecology therefore does not include the full prohibition of any chemical input. But in each and any case, they should only be used as a last resort and at the lowest level possible. The agroecological approach clearly requires reducing off-farm inputs (chemical or biological) to an absolute minimum. Besides, it necessarily excludes any use of genetically modified organisms (GMOs). Many reasons explain why GMOs are incompatible with agroecological farming.

Some of the main reasons why agroecology and gmos are incompatible.

The development of GMOs presents potential or proven risks including the following:

- Increased peasants' dependence on the agro-industry and thus reduced autonomy of farmers (notably by prohibiting farmers' to save seeds themselves);

- Biodiversity reduction (weakening flexibility offered by the natural environment to design adequate context-specific agroecological strategies);

- Harmful impacts on the environment (e.g. through adverse impacts on beneficial insects and other organisms); increased environmental threats to farming systems (e.g. through the development of secondary pests resistance);

- Increased vulnerability of farming systems (notably due to biodiversity reduction);

- Reduced natural soil fertility;

- Increased economic costs for peasants and restricted experimentation by individual farmers while potentially undermining local practices for securing food and economic sustainability;

- Increasing criminalization of peasants linked to the development of patents and the context of unfavorable national seeds laws and legislations, as illustrated in recent years by Monsanto practices in Northern America.

Relying first and foremost on traditional knowledge does not mean excluding modern science. In fact, agroecology combines scientific inquiry with indigenous knowledge, as well as farmers' innovation and community-based innovation for shaping sustainable farming systems. For instance, in Central America the coffee groves grown under high-canopy trees were improved by the identification of the optimal shade conditions, minimizing the entire pest complex and maximizing the beneficial microflora and fauna while maximizing yield and coffee quality. Generally speaking, the role of agronomists and other researchers is very important for making agriculture more agroecological, not only for contributing significantly to agroecological innovations, but also for helping better understand and address global sustainability challenges beyond the farm, at the territorial level.

Moreover, agroecology should not be seen as incompatible with the mechanization of agriculture. While a forced path toward a rapid mechanization of farming that does not meet peasants needs should be avoided, agroecological farming is perfectly compatible with a gradual and adequate mechanization of farming. One illustration is provided by the in-depth analysis of the evolution of the agrarian systems of the Nile Valley, which has shown a successful adaptation of mechanization to the size and needs of these peasant farming systems, with most of the soil preparation work and water pumping and gain threshing being mechanized. The small scale of plots is not an obstacle, for example, to mechanized water pumping because water is brought by gravity to the third level canals where it is usually pumped and brought to private canals running along the land parcels. This in-depth analysis has shown that decent living conditions could be reached for a family with a plot of good land of a size between 0.5 and 0.8 ha, with the appropriate mechanization and animal-crop integrated systems.

One fundamental feature of agroecology as a holistic agricultural transition process is the systematic search for the best combinations of techniques and strategies, instead of relying on a few standardized best practices, for optimizing sustainability performances of farming systems. The challenge is to identify the most efficient socio-technical arrangements in heterogeneous environments, the right combinations of practices that will best allow for realizing agroecological principles. Those combinations will necessarily vary from one context to another, since each context has its own characteristics and therefore its own conditions to achieve sustainability. This is one of the reasons why, while some types of practices have been typically described as agroecological, agroecological farming cannot be reduced to a 'catalogue of techniques' whose standardized application would automatically bring sustainability. Agricultural sustainability does not depend on the intrinsic characteristics of a few magic bullet solutions that would be independent from the environment to which they apply. It relies on the quality of complex interactions that result from an entire package, adequate combination of various practices whose operationalization in particular circumstances will necessarily have to change depending on each context.

Types of Practices Typically Promoted as Agroecological

Jules Pretty, from University of Essex in the United Kingdom (UK), has highlighted seven agroecological practices and resource-conserving technologies:

1. Integrated pest management (IPM), which uses ecosystem resilience and diversity for pest, disease and weed control, and seeks only to use pesticides when other options are ineffective.

2. Integrated nutrient management, which seeks both to balance the need to fix

nitrogen within farm systems with the need to import inorganic and organic sources of nutrients, and to reduce nutrient losses through erosion control.

3. Conservation tillage, which reduces the amount of tillage, sometimes to zero, so that soil can be conserved and available moisture used more efficiently.

4. Agroforestry, which incorporates multifunctional trees into agricultural systems, and collective management of nearby forest resources.

5. Aquaculture, which incorporates fish, shrimps and other aquatic resources into farm systems, such as into irrigated rice fields and fishponds, and so leads to increases in protein production.

6. Water harvesting in dry land areas, which can mean formerly abandoned and degraded lands can be cultivated and additional crops grown on small patches of irrigated land owing to better rainwater retention.

7. Livestock integration into farming systems, such as dairy cattle, pigs, and poultry, including using zero-grazing cut and carry systems.

Depending on how it is concretely applied and completed or not by other practices, one particular technique can sometimes either be an active component of a truly agroecological farming system, or on the contrary contribute to non-sustainable externalities. This can be well illustrated with no-till. Also referred to as 'zero till', 'no-till' is usually defined as "a system of planting (seeding) crops into untilled soil by opening a narrow slot, trench or band only of sufficient width and depth to obtain proper seed coverage. No other soil tillage is done". Detailed scientific evidence exists showing that no-till conserves the natural resources in the soil and water through various mechanisms. Decreases in soil erosion and water losses are often spectacular and are reported from many sites. But since tillage impacts include some weed, pest, nutrient or water management effects, if a farmer abolishes tillage without changing anything else in the cropping system, this will induce in most cases problems with weeds, pests and nutrient availability and might require more herbicides, pesticides and fertilizers. No-till can therefore easily be one component of industrial farming systems. As a matter of fact, scientific sources and statistics indicate that no-till today often comes 'in a package' with monocultures, GMOs and extensive herbicide use.

But no-till can also be combined with natural control mechanisms for managing insect pests, pathogens and weeds and therefore reducing the need of further artificial interventions. For example, in Santa Catarina, southern Brazil, many hillside family farmers have modified the conventional no-till system. Instead of relying on herbicides for weed control, these innovative organic minimum tillage systems rely on the use of mixtures of summer and winter cover crops which leave a thick residue mulch layer, on which after the cover crops are rolled, traditional grain crops (corn, beans, wheat, onions, tomatoes, etc.) are directly sowed or planted. Depending on the cover crop or cover crop combination used, residues have the potential to suppress weeds. But weeds' response to residue depends on various factors, such as the type, quantity and thickness of residue applied, the time remaining as effective mulch, cover crops used and biology of particular weed species. Experience shows that simply copying the cover crop mixtures used by successful farmers won't work for widely diffusing the technology. Agroecological performance does not depend on specific species or techniques, but is linked to processes optimized by the whole system. Optimization consists in increasing the degree of 'agroecological integration', that is the extent to which a given farming system realizes agroecological principles. Assessing the sustainability of a given farm hence can be seen as consisting in assessing its degree of agroecological integration, ranging from an industrial monoculture (negligible agroecological integration), to a monoculture-based organic farm with input substitution (low level of integration), to complex peasant agroforestry system with multiple annual crops and trees, animals, rotational schemes, and perhaps even a fish pond where pond mud is collected to be used as an additional crop fertilizer (high level of agroecological integration).

Applying a Bottom-up and Farmer-led Approach

While the Green Revolution model has favored a top-down approach which tends to reduce peasants to no-choice passive recipients of technology received from extension agents or inputs suppliers, agroecological transition requires bottom-up processes in which farmers take the front seat. Conventional top-down extension can be demobilizing for farmers, as technical experts have all too often had the objective of replacing peasant knowledge with purchased chemical inputs, seeds and machinery. On the contrary, agroecological farming is highly knowledge-intensive and based on techniques that are not delivered top-down but developed on the basis on farmer's knowledge, experimentation and innovation.

Different methodologies have been developed for promoting farmer innovation and horizontal sharing and learning. The Campesino-a-Campesino (farmer-tofarmer, or peasant-to-peasant) methodology (CaC) is one of the most often used. CaC is a Freirian horizontal communication methodology, or social process methodology, that is based on farmer-promoters having developed new solutions to problems that are common among many farmers or have recovered/rediscovered older traditional solutions, and who use their own farms as their classrooms to share them with their peers. Based on local peasant needs, culture and environmental conditions, CaC is mobilizing because it makes peasants the protagonists in their own processes of generating and sharing their own (and appropriated) technologies. Another method is the Farmers Field Schools (FFS) approach that has been developed and promoted by FAO as part of its ecological approach called Integrated Pest Management (IPM) in South East Asia. In this group-based discovery-learning process, farmers observe, record, and discuss what is happening in their own fields instead of listening to lectures or watching demonstrations. The process generates deep understanding of farming problems and promotes practical communication mechanisms for its solution.

Agroecology as a Movement

As we have seen, since the 1970s the concept of agroecology has also refered to a number of collective mobilizations, originally in response to the Green Revolution. Agroecology as a movement has been particularly strengthened politically in the last 5 years through LVC, the largest transnational peasant movement, as one of the key pillars of Food Sovereignty. It is also politically supported by other farmers' umbrella organizations and peasant movements, sometimes but not always members of LVC, such as the East and Southern African Farmers' Forum (ESAFF), the Network of Farmers' and Agricultural Producers' Organisations of West Africa (ROPPA – Réseau des organisations Paysannes et de producteurs de l' Afrique de l'Ouest), the Landless Workers' Movement (MST – Mouvement des Sans-Terre) in Brazil or Bolivia, or the Latin American Coordination for Peasant Organisation (CLOC), an umbrella organisation with 84 sub-organisations in 18 Latin American and Caribbean countries.

The concept of Food Sovereignty was first used by LVC on the international scene in 1996 during the World Food Summit held in Rome, and has been further elaborated on at the International Forum for Food Sovereignty hosted in 2007 by LVC in Nyéléni, Mali, to which LVC invited sister international movements of indigenous people, fisher folk, women, environmentalists, scholars, consumers and trade unions. Its core definition developed on that occasion defines it as "the right of peoples to healthy and culturally appropriate food produced through ecologically sound and sustainable methods, and their right to define their own food and agriculture systems". Among others, this implies "the rights to use and manage lands, territories, waters, seeds, livestock and biodiversity are in the hands of those of us who produce food", as well as "the rights of consumers to control their food and nutrition".

The inclusion of agroecology in the broader framework of Food Sovereignty is therefore not surprising. Indeed, as an attempt to protect peasant agricultures from the growing pressure of industrial agriculture, the whole point of agroecological farming is precisely to achieve sustainable agriculture for meeting local needs through ways that enhance the autonomy and control of peasants over their own production systems, instead of making them more dependent on off-farm inputs and external experts. In that sense, agroecology appears as a key strategy of what van der Ploeg calls re-peasantization, a concept that not only refers to a growing number of peasants (quantitative dimension), but also entails a qualitative shift consisting in people entering the 'peasant condition', characterized by the constant search of an increased autonomy. When farmers undergo a transition from input-dependent farming to agroecology based on local resources, they are becoming 'more peasants' since they are gaining autonomy. To some extent, transitioning existing peasant systems towards agroecological farming could be seen as a process leading to the full realization of the 'peasant logic'.

The search for autonomy can also rely on the development of alternative agrifood networks (AAFNs) such as producer–consumer networks, collective producer shops, farmers' markets, box schemes and school provisioning schemes. Just as the industrial agri-food system supports industrial agriculture and opposes attempts to shift it towards sustainable agriculture, AAFNs are frequently supportive of and rooted in agroecological farming, and seek to decrease reliance on industrialized agri-food systems. They work against the logic of bulk (high volume and low cost) commodity production, redistribute value through the food chain, rebuild trust between producers and consumers, and articulate new forms of political association and market governance.

For LVC it is clear: agroecology cannot be reduced to its technical ecological content but also encompasses social and political dimensions. It politicizes what used to be seen as purely technical questions of farming. It opposes the industrial agricultural and food, capitalist rural development model, giving to agroecological transition processes emancipatory potential. This understanding of agroecology has led LVC to strive politically for its scaling-up. LVC has been struggling for scaling-up agroecology by denouncing agrofuels, GMOs, carbon markets, REDD and REDD+ as 'false solutions' to climate change, and by stressing publicly the risk of cooptation of agroecology through the paradigm of sustainable intensification. Striving for scaling-up agroecology consists both in advocating for policy measures and regulations specifically supportive of agroecology, and in challenging the obstacles, resulting from a range of various policies and economic practices (e.g. trade and agricultural liberalization), that have historically disadvantaged peasant agricultures in many national, regional and international contexts. Addressing those obstacles is needed to unleash the tremendous sustainability potential that peasant agricultures traditionally hold (as demonstrated by agroecology as a science), a potential which then, through an agroecological modernization process, can be strongly increased by combining traditional knowledge and know-how with the best available modern agroecological science.

But the importance of advocating politically for defending and scaling-up agroecology is not carried by all civil society actors. Historical divisions exist between farmer-to-farmer and NGO-based networks whose work has concentrated on promoting the adoption of agroecological farming to more farmers (horizontal scaling-up), and agrarian-based farmer organizations and movements such as LVC who have engaged politically (vertical scaling-up). Farmer-to-farmer and NGO-based agroecology networks, such as the Farmer to Farmer Movement (CAC - El Movimiento Campesino a Campesino) active in a dozen countries of Latin America, or the Participatory Ecological Land Use Management (PELUM), a regional network of over 207 civil society organisations which operate in 10 African countries (Botswana, Kenya, Lesotho, Malawi, Rwanda, South Africa, Tanzania, Uganda, Zambia and Zimbabwe), have been highly effective in supporting local projects and developing sustainable practices on the ground. On the other hand, unlike LVC, they have done relatively little to address the need for an enabling policy context for sustainable agriculture. For political advocates, these practitioners have historically tended to reduce agroecology to technical and apolitical approaches to agricultural development. This has led advocates to call many NGOs to shift from technology-led agendas to strategies that support farmerled political organizations.

Progress in the agenda for Food Sovereignty however is being made. Convergence is growing progressively between practitioners and advocates. Slowly but surely, distinct groups begin to see themselves as part of a larger movement to develop civil society. For example, as a result of members of PELUM willing to engage in more agrarian advocacy the ESAFF was formed in 2002, as a farmer's voice in East and Southern Africa. PELUM and ESAFF work closely together, with ESAFF challenging PELUM on a number of issues. Such evolution and many others suggest that the international struggle for Food Sovereignty, as understood by LVC, is beginning to take root in smallholder agroecology networks. Similarly, LVC enhances its efforts to spread agroecological approaches throughout its own farmer organizations, which remains a challenge.

Is Sustainable Intensification of Agriculture a Better Path?

While agroecology has been subject to increased worldwide attention and scrutiny in the last few years given the growing awareness of current agricultural and food sustainability crisis, the most influential actors in the debate prefer advocating for a 'sustainable intensification of agriculture'. These include governments of the USA, the EU and UK, FAO, IFAD, the World Bank, research institutions and centers including the Consultative Group on International Agricultural Research (CGIAR) and its 15 research centers, as well as agribusiness companies and organizations such as the Agricultural Biotechnology Council and the International Fertilizer Industry Association, or the Bill and Melinda Gates Foundation. At first sight, using one or the other of those concepts might seem rather anecdotal, especially since in the way it is being used by these actors, the concept of 'sustainable intensification of agriculture' does include agroecological practices. Looking at it more closely, one can understand how privileging this term (rather than advocating more directly for scaling-up agroecology) is far from being anodyne.

Promoted as a solution for small farmers in developing countries, sustainable intensification is presented as a step change in agricultural science and development, re-conciliating sustainable agriculture with intensive farming, creating an environmentally benign agriculture that also improves yields. For example, the Royal Society defines the challenge of sustainable intensification as intensification "in which yields are increased without adverse environmental impact and without the cultivation of more land". This sounds close to agroecological farming. Do agroecological approaches not enable farmers to enhance yields sustainably (as shown in Part II)? But this is only an appearance. Indeed, rather than calling for a radical shift of agricultural development, the sustainable intensification agenda is a reformist one, complementing conventional approaches inherited from the Green Revolution model by a more systemic approach to sustainably managing natural resources, including through a more selective use of external inputs. It aims to offer an inclusive and flexible menu in which 'no techniques or technologies should be left out', but should be combined to each other depending on the context. And here's where things start to get complicated: GMOs are promoted as part of the solution, along with conventional practices and agroecological practices.

In its Reaping the benefits report published in 2009, The Royal Society has well explained the rationale for such 'inclusive' logic: "Past debates about agricultural technology have tended to involve different parties arguing for either advanced biotechnology including GM, improved conventional agricultural practice or low-input methods. We do not consider that these approaches are mutually exclusive: improvements to all systems require high-quality science. Global food insecurity is the product of a set of interrelated local problems of food production and consumption. The diversity of these problems needs to be reflected in the diversity of scientific approaches used to tackle them. Rather than focusing on particular scientific tools and techniques, the approaches should be evaluated in terms of their outcomes. Recent progress in science means that yield increases can be achieved by both crop genetics (using conventional breeding and molecular GM) and crop management practices (using agronomic and agroecological methods)". The Royal Society promotes GMOs as a potential option notably for increasing farmers resilience to climate change (e.g. through the use of drought tolerance crops) and pests attacks (through herbicidetolerant seeds), or for improving food nutritional quality (e.g. with 'golden rice' for combating vitamin A deficiency). Other influential actors mentioned above also typically promote GMOs as a potential solution when advocating for a sustainable intensification of agriculture. According to IFAD, for example, second generation of transgenic crops designed to perform well under drought, flood, heat and salinity "may play a greater role in addressing this set of issues, which can greatly contribute to reducing the risks faced by smallholder farmers".

The consideration of GMOs as part of the solution is highly problematic, since it is simply incompatible with a truly agroecological development paradigm for obvious reasons. But the true challenge for better understanding what agricultural development models these actors are concretely supporting, is looking beyond the rhetoric on how their funds are spent when investing in 'sustainable intensification'. The NGO Friends of the Earth International (FoE) has recently made a helpful contribution to such monitoring. Based on existing evidence, its October 2012 report A Wolf in Sheep's Clothing? An analysis of the 'sustainable intensification' of agriculture provides useful information on funding priorities of the UK Government, the Bill and Melinda Gates Foundation, the CGIAR and the US Government in terms of sustainable intensification agricultural research and development projects. It notably concludes that while claiming to include agroecological farming, the sustainable intensification agenda in practice seems to focus primarily on technology-based approaches including GMOs, further consolidating industrial agriculture. The Feed and Future agricultural development programme of the US Government, launched in 2009 and led by the US Agency for International Development (USAID), provides a good illustration. Relying on the philosophy of sustainable intensification that it defines as being close to conventional intensive agriculture, its research strategy notably includes developing drought and stress tolerant crops, disease and pest resistant crops, crops with improved nitrogen use efficiency and yield improvements. When USAID staff gave an outline of funding priorities in 2011, they revealed that 28% of research funding would be directed to 'climate resilient cereals'. The Feed and Future programme priorities in target countries also encourage the adoption of Conservation Agriculture, which in USAID's vision consists mainly in no-till farming systems completed with high levels of chemical inputs and often use GM crops that don't require tilling for weed control.

The GM Freeze campaign, whose members include various NGOs such as FoE England, GeneWatch UK, EcoNexus and FARM, raised similar concerns with regard to the Gates Foundation. According to the campaign, the Gates Foundation has allocated more than eight times as much money to the Alliance for a Green Revolution in Africa (AGRA) for a project to distribute artificial fertilizers as its main activity than to researching improved soil fertility using local resources, and funding for research involving transgenics outstrips that for soils by more than ten-fold. Invariably, in the framework of the sustainable intensification agenda, agroecology receives a fraction of the funding provided to Green Revolution technologies.

But the sustainable intensification agenda does not just give only a small amount of available funds to agroecology when investing in agriculture. It also reduces it to its ecological technical content, essentially ignoring its social and political dimensions. In that sense, agroecology can be seen as co-opted by actors who fundamentally do not want to question the prevailing system (since their objective interests depend on it) but rather seek to proceed to the minimum adjustments that are necessary for ensuring the reproduction of the dominant industrial, corporate food regime. Agroecology then becomes a means (rather than a barrier) for the expansion of industrial agriculture.

Agroecological Principles Applied to Large-scale Industrial Agriculture

The close relationship between agroecology and peasant agricultures is obvious: agroecological systems are deeply rooted in the ecological rationale of traditional small-scale agriculture. Modernizing agroecologically traditional small-scale farms is thus especially appropriate for improving significantly their sustainability performances, notably for boosting yields and productivity per unit of land. This is good news for traditional peasants who do not use a tractor, working animal, selected purchased seeds, mineral fertilizers, or pesticides. According to Mazoyer, the number of such peasants would amount to roughly 500 million people (of a total active agricultural population estimated to 1,34 billion people).

Most industrially 'accomplished' small-scale farmers, increasing the agroecological integration of their farms will be more difficult. Indeed, the conversion of degraded, simplified production systems to diverse, agroecological, resilient, low carbon systems, is challenging. The challenge will notably consist in avoiding excessive decline of yields and land productivity that would result from a too sudden abandon of synthetic inputs. In such cases, it can take time before beginning to recover and build productivity again, through the restoration of local ecosystems health. As a consequence, the transition processes will need to be more progressive19. However, shifting those farms into agro-ecological systems remains technically very much possible.

With regard to the technical feasibility of agroecological transition processes to various agricultural systems, the real challenge concerns large-scale industrial farms. To what extent can agroecological principles be applied to those farms? Is it realistic? Very few references in the agroecological literature provide elements to answer this question. Among them, Altieri et al. seem to answer positively, underlining that in countries such as Chile, Argentina and Brazil, large plantations are now being re-thought with a paradigm based on circular systems with reduced input and energy consumption rather than focusing solely on linear approaches and on increasing throughput. They posit that although the diversity of crops and the integration animal-crop may be less obvious than it is on small plantations, the same overall principles apply. On the other hand, few authors stress the limitations of attempts to increase the degree of agroecological integration of large industrial farms. For example, Lin implicitly emphasizes how unsuitable to biodiverse farming systems industrial mechanization is, since it is designed for optimizing productivity for one crop type and one crop structure. Douillet and Girard write along the same lines when emphasizing that the standardization of cropping systems has precisely promoted industrial mechanization.

Though it is hard to provide a comprehensive answer to the question, logical conjectures suggest that in most cases agroecological integration of large industrial farms can be increased, but that room for maneuver is necessarily limited. For example, it is difficult to imagine how exactly large farms managed by just one or at best a few people could adopt farming management systems that entail enhancing significantly on-farm biodiversity, or total output per ha, to the same extent as peasants do on small plots of land. However, this remains a hypothesis to be tested.

Whether fully applying agroecological principles to large industrial farms is technically possible or not is an important question, since it gives us indications as to the feasibility of transitioning from industrial farming systems towards truly more sustainable farms. Answering 'no' would imply that above a certain size, sustainability of agriculture will necessarily be restricted. The question is relevant. As a matter of fact, there is an ongoing debate on the nature of relationship between farm size and productivity of outputs like crop yields and biodiversity.

Does it mean that large farms should be converted into smaller farms? Not necessarily. It should be so in countries those are highly dependent on agriculture and where peasants and communities are suffering from an inequitable access and control over land and other natural resources due to an unfair competition with large industrial farms. In such contexts there is no justification, from a social equity perspective, for not fragmenting large farms into smaller units through adequate redistributive land reforms. By contrast, since agroecological farming is labor intensive, any attempt to promote significantly smaller farms and making them agroecological in areas of very low population density or where too few people want to work in agriculture, and in which peasants do not suffer from such inequitable access and control over land and other natural resources, would not make sense. In such regions, increasing the agroecological integration of large industrial farms to the extent possible may be the best option for improving agricultural sustainability, through adequate incentives, both positive and negative (for encouraging the best and discouraging the worse practices respectively). In particular, in such areas the adoption of LEI (low-external-inputs) agriculture practices by large-scale farming will be crucial to mitigate adverse environmental impacts.

This is especially the case in many industrialized countries. Hendrickson et al. for example emphasize demographics as one of the three key factors limiting the adoption of integrated farming in the US, and the problem is also widely recognized in the EU. Still with regard to the US and the EU, Wibbelmann et al. note that "the trend of rural depopulation has a powerful effect on the human capital needed to increase the adoption of agroecological approaches, and this is exacerbated by low agricultural wages which are not conducive to labour movements into rural areas".

However, constraints imposed by the size of the farms are far from applying everywhere in the developed world. In Europe for example, the average surface area used per farm varies considerably from one region to another. In 2010, this average reached 14,1 ha in the EU-27, varying from 0,9 ha in Malta to 152 ha in the Czech Republic, with 7 Member-States (Romania, Italy, Poland, Spain, Greece, Hungary and France) accounting for more than 80% of the European farms. It should also be noted that in Europe land grabbing is a reality as well, including for agricultural purposes, as documented in a joint, comprehensive publication launched in April 2013 by ECVC and the Hands off the Land network. Among other issues, young people wishing to set up farming are facing major barriers to land ownership and access, including increasing costs of agricultural land. These elements indicate that in many European countries the room for agroecological transitions is real. Moreover, in some of these countries, the continuing decline of agricultural jobs is put into question in the name of non-market functions of agriculture, such as land occupancy, or due to enthusiasm for short circuits, connecting urban citizens and producers.

Challenges for Scaling-up Agroecological Approaches

Contrary to what its detractors often claim, agroecology is scalable. As a matter of fact, it has already been spread and applied by many farming communities worldwide, reaching millions of farmers and millions of ha in Africa, Asia and the Americas, as documented by several major global assessments. Yet, it could be far more diffused, and given the great potential it offers for meeting sustainability challenges, it should. Disseminating agroecological farming means first promoting its adoption by more farmers through farmer-to-farmer networks ('horizontal scaling-up', also referred to as 'scaling-out'). But ensuring its adoption to a significantly higher stage will also and crucially require institutionalizing supportive policies ('vertical scalingup'), breaking with cycles of policies which all too often have disadvantaged peasant agricultures and agroecology, such as mainstream trade and agricultural policies including the structural adjustments programs of the International Monetary Fund (IMF) and the World Bank, and the Agreement on Agriculture of the World Trade Organization (WTO), and with the current trends in agricultural reinvestments which tends to consolidate industrial agriculture through the reformist agenda of sustainable intensification. Experience shows that with adequate support and investment from the State, agroecology can be efficiently scaled-up at a higher level44. This requires political will and, ultimately, a real democratization of agricultural and food governance.

Unlocking Ideological Barriers to Political Recognition

Generally speaking, there is a need to enhance the recognition among key decision makers of agroecology and its benefits in achieving sustainable agricultural and food systems. As long as they are not convinced of these benefits, it is unlikely that they will create the enabling institutional environment that is needed for prioritizing its scaling-up in agricultural development. But such political recognition is impeded by persisting misconceptions about agroecology and peasant agricultures.

For example, when discussing and negotiating with governments' representatives within the CFS, one easily notes that the majority still perceive agroecology as one particular set of predetermined practices only adaptable to very few, limited contexts. Another example is the persisting belief that monocultures and industrially managed systems, or large farms are at all levels more productive than diversified small-size agricultural systems. Furthermore, agroecology is often mischaracterized as a 'return to the past' or a model incompatible with a (gradual) mechanization of agriculture, as if the only choice we had for developing agriculture was between 'modern' (industrialized) farming and traditional (archaic) peasant agricultures and that optimizing sustainability performances of these traditional forms of agriculture through agroecology could not be regarded as modernizing them. These cultural perceptions are sometimes so prevalent that even for many smallholders themselves, the use of industrial technologies such as synthetic fertilizers, pesticides, transgenic and hybrid plant varieties, or mono-cropping may be a lever of social integration if farmers can openly demonstrate their capacity to afford them, as illustrated in the Indian context in which traditional farming systems are often stigmatized as an anachronism. Furthermore, performance criteria used to monitor agricultural projects are most often still narrowly limited to classical agronomical criteria and measures such as yield and productivity per unit of labor, instead of complementing them by comprehensive indicators better able to measure sustainability, including for example the productivity of land or water, the impacts of agricultural projects or technologies on incomes, resource efficiency, hunger and malnutrition, empowerment of women and other beneficiaries, ecosystem health, public health and nutritional adequacy. Such narrow a priori forms of perception constitute a major obstacle for building markets and economies that take into account social and environment costs.

Reversing these misconceptions is a not sufficient but necessary condition for enhancing the political will to prioritize the scaling-up of agroecological approaches in agricultural development. This will require first and foremost efforts in awareness raising and dissemination among relevant key decision makers, extension agents and farmers organizations. Among other objectives, those efforts should seek to stress the economic viability of agroecological farming. Indeed, in order to engage in an agroecological transition process, farmers need to be sure this represents an economically viable option. But many of them lack enough information about profitability and tend to fear, on the contrary, economic losses.

Supporting Farmer-to-Farmer Networks

Given the bottom-up approaches that agroecology implies, using and building upon the resources already available (local people, their knowledge and their domestic natural resources), its scaling-up requires inclusive, community-oriented methods for networking and sharing techniques. Localized farmer-to-farmer networks are crucial in the dissemination of information between farmers in similar agroecological zones. Successful scaling-up relies heavily on enhancing human capital and empowering communities through training and participatory methods that seriously take into account the peasants' needs, aspirations and circumstances. Highly organized peasant organizations are extremely important in this regard. Farmer-to-farmer networks and organized social rural movements such as LVC, compromising around 150 local and national organizations in about 70 countries, the one million families MST in Brazil, or ANAP in Cuba must therefore be actively encouraged and supported. Farmers' organizations and networks have accumulated experience on the dissemination of agroecological practices in the last decade, with proven results. They are functioning as learning organizations and must be supported in this role. In Cuba, in just ten years' time, ANAP has been able to build a grassroots movement for agroecology leading to spread to more than one third of all peasant families the transformation of productions systems into agroecological and diversified farming systems.

Training farmers and disseminating best approaches for transitioning agricultural systems towards agroecological farming happens through a great variety of participatory methods such as field days, on-farm demonstrations, trainings of trainers or farmers' cross-visits. FFS have been very efficient in empowering peasants by helping them to organize themselves better, and stimulating continuous learning. The successful dissemination of the push-pull strategy (PPS) in East Africa by the International Centre for Insect Physiology and Ecology (ICIPE) provides a good example46. It is largely due to the demonstration of fields managed by model farmers, receiving visits by other farmers during field days, as well as to partnerships with national research systems in Tanzania, Uganda, Ethiopia and other countries.

But peasants' organizations and farmer-to-farmer networks are not only important for disseminating agroecological farming. They also enhance farmers' skills to advocate for their rights and Food Sovereignty. LVC has created political leadership training academies in many countries and regions for preparing peasant leaders to pressure public authorities at the local, provincial, national and international level to obtain more alternative, more agroecology-, climate-, farmerand consumer-friendly public policies, including by organizing massive mobilizations when confronted with less friendly policy makers. LVC has also been struggling for scaling-up agroecology by denouncing agrofuels, GMOs, carbon markets, REDD and REDD+ as 'false solutions' to climate change, and by stressing publicly the risk of cooptation of agroecology through the paradigm of sustainable intensification.

Despite successes achieved so far, the lack of appropriate social networks for allowing collective experimentation and exchange of information by peasants over agroecology remains an important constraint that limits its dissemination and adoption by farmers at a higher stage. The establishing and functioning of such networks needs adequate support. Moreover, increased collaboration and coordination is needed among the various actors (farmers' organizations, public authorities, NGOs, academic institutions and research centers) for boosting scaling-up efforts.

Providing an Enabling Public Policy Environment

Unlocking ideological barriers to its political recognition and supporting localized farmer-to-farmer networks will not be sufficient for scaling-up agroecological approaches at a higher level. On the longer term, it will also be crucial that public authorities ensure a favourable environment to promote it. Since agroecological systems are deeply rooted in the ecological rationale of traditional small-scale agriculture, such an environment will primarily consist in implementing supportive policies to peasants in general (progressively dismantling the policies that have historically hindered their development) and adopting specific incentives to modernize agroecologically both the most traditional peasant agricultures and those partially industrialized. However, specific action will also be needed for addressing as much as possible the non-sustainability of large-scale industrial farms. In synthesis, action will be required at the following four levels:

1. Designing agricultural and trade policies in support of peasants and agroecological approaches;

2. Securing peasant's access to natural and other productive resources;

3. Supplying public goods;

4. Prioritizing agroecology in agricultural research and extension services.

1. Designing Agricultural and Trade Policies in Support of Peasants and Agroecological Approaches:

Most often, the dynamic of today's agricultural and markets policies, both at the domestic and international levels, seriously undermines peasant agricultures and are a great constraint for scaling-up agroecology. Peasants are fully part of different markets but their position in these markets is weak. Some of the main unfavourable conditions include: low commodity prices and food prices volatility, further reducing net incomes in a context of high costs of inputs and/or increased food prices (peasants suffering from such increases as consumers while being badly affected as producers by low commodity prices); lack of power and negotiation capacity of most small-scale farmers within the agrifood value chains; cheap imports reducing domestic peasants' outlets on local markets; non internalization of environmental and social costs in agricultural and food prices; downsizing of public services and disinvestment in agricultural systems; increased fierce international competition of peasant agricultures characterized by huge competitiveness gaps.

Such unfavourable conditions result from various adverse policy and economic factors such as: the liberalization of agricultural trade, including through the structural adjustment programs of the IMF and World Bank in the 1980s and 1990s and the Agreement on Agriculture of the WTO which have among other things significantly contributed to import surges in developing countries; the high concentration in agrifood value chains which tend to be increasingly dominated by a few large corporations controlling the distribution channels between farmers and consumers, and which among other things reduces to the smallest proportion the gains received by peasants of the final prices of food products; the internationalization of value chains and the rise of supermarket and buyer-driven chains whose impacts notably include increasing difficulties for small-scale producers to meet volumes and standards requirements of global buyers and retailers; agroexports and export dumping policies primarily based on a lack of adequate supply management policies in export countries as illustrated by the US Farm Bill and the European Common Agricultural Policy (CAP); or agricultural subsidies policies advantaging monocropping and corollary discouraging biodiverse farming systems.

On the contrary, public authorities should use all available policy tools to make markets better work for peasants and agroecological approaches. The challenge is both to regulate differently and better the existing markets, as well as developing new ones. Consumers could contribute to this shift by pressuring governments and economic actors to take action.

2. Securing Peasants' Access to Natural and other Productive Resources:

Ensuring the rights of farmers to access, breed, produce, conserve, purchase, exchange and use the seeds they need is of utmost importance from an agroecological perspective. When peasants' access to and control over seeds are threatened, their flexibility to design sustainable farming systems that are adapted to their particular needs and to the specificities of each local context, is undermined. It is therefore not surprising that for LVC, access and control over seeds is the very basis of Food Sovereignty.

Peasants' adequate access to and control over land, water and other natural resources is also essential, firstly because peasants need to be able to mobilize resources to manage agroecological strategies and practices. In that sense, scaling-up agroecological approaches at a higher stage implies ensuring a responsible governance of tenure of land, water and other natural resources. Among other things, such responsible governance includes tackling land concentration and ensuring a fair share of land, water and other natural resources, for allowing every household to mobilize these resources on a relatively smallscale that is typical of peasant agricultures. Improved security of tenure is also important for encouraging farmers to invest in the long-term sustainability of the environment (e.g. through the planting of trees, the more responsible use of soils and other practices with long-term payoffs), since they will be more motivated to take care of the land and other natural resources their livelihoods depend on if they can be ensured they won't lose them to industrial or urban developers of large scale agricultural business. Water and other natural resources allows (re)conciliating in a sustainable way various actors' (potentially conflicting) expectations as to the use of these resources, which is important for successful agroecological transitions.

Yet, a large-fraction of peasants and many urban poor are persistently suffering from a lack of adequate and secure access and tenure over land and other natural resources, which is one of the main causes of hunger and poverty in the world. Furthermore, this trend has recently increased with the phenomenal acceleration of large-scale land acquisitions following the agriculture and food prices spike in 2008. Land, water and other natural resources are being grabbed to serve various commercial interests and purposes, such as the massive production of biofuels. Access to seeds and in particular traditional varieties is increasingly threatened by the extension of 'modern' varieties, mainly hybrids, and the expansion of IP Rights regimes in agriculture, which tend to create a market for seeds dominated by few large companies and do not provide any incentives for in situ conservation by farmers since they do not reward their role in conserving and improving landraces for breeding.

This must change. Peasants' rights to seeds, as well as to land, water and other natural resources should be respected and protected. Moreover, although agroecological farming reduces the need for credits, peasants must also benefit from an improved access to them when needed for investing in their own development.

3. Supplying Public Goods:

Scaling-up agroecological approaches requires the supply of public goods such as rural infrastructure (roads, electricity, information – including up-to-date information on commodity prices – and communication technologies, irrigation systems) and therefore access to local and regional markets, access to credits and insurance against weather-related risks, agricultural research and extension services, storage and handling facilities to reduce postharvest losses in rural areas, education and sanitation.

Reaffecting part of public spending on private goods (such as fertilizers or pesticides that farmers can only afford as long as they are subsidized) to public goods can bring significant sustainability benefits. This can be illustrated through a research based on the study of 15 Latin American countries over the period 1985- 2001, which has shown that within a fixed national budget, a reallocation of 10 % of spending on private goods to supplying public goods increases per capita agricultural income by 5 %, while a 10 % increase in public spending on agriculture, keeping the spending composition constant, increases per capita agricultural income by only 2 %. A World Bank Policy Research Working Paper concluded that "even without changing overall expenditures, governments can improve the economic performance of their agricultural sectors by devoting a greater share of those expenditures to social services and public goods instead of non-social subsidies".

The primary responsibility of States in ensuring the supply of public goods is particularly obvious given the general lack of natural incentives for private companies to invest in these domains, and transaction costs that are too high for local communities to create these goods themselves.

4. Prioritizing Agroecology in Agricultural Research and Extension Services:

Reinvestment efforts in agriculture since 2008 have essentially lead to the further expansion of a 'somewhat-less-polluting' industrial agriculture, while agroecological approaches have been poorly supported. This trend notably applies to agricultural research and extension services. Public agricultural research and extension agents are increasingly being influenced by private interests to promote conventional approaches rather than agroecology. Action Aid's fieldwork in Guanzi region in China, for example, found that extension services are poorly encouraging sustainable agriculture and instead are vigorously promoting hybrid seeds, pesticides and fertilisers. The recognition of the growing challenges that agriculture will have to face as a consequence of climate change has brought about a major effort to adapt agriculture through technical means, primarily the research and development of drought-resistant biotech crops. Clearly, much of today's publically funded research does not meet the needs or priorities of peasants in low- and middle-income countries. In West Africa, the agricultural research system, which relies heavily on external funding, has developed genetically improved varieties of sorghum, millet or groundnuts which tend to be hybrids and therefore cannot be resown year after year, while often also requiring additions of chemical fertilisers and pesticides, thus increasing farmers' dependence on purchasing and their risk of debt.

The "Democratising the Governance of Food Systems. Citizens Rethinking Food and Agricultural Research for the Public Good" international action-research initiative provides a prime example of the yawning gap between peasants' priorities and the mainstream further expansion of the agro-industrial model in agricultural research and extension services. Started in 2007 on the initiative of IIED and local partners, the project uses participatory methods and institutional innovations to create inclusive, democratic and safe spaces for citizens to get involved in research policymaking and agenda setting in four regions, with one country acting as host for each region: West Africa (Mali), South Asia (India), West Asia (Iran) and the Andean region in Latin America (Bolivia). In Mali, as a first step in 2009, African partners, Biodiversité: Échanges et Diffusion d'Expériences (BEDE) and IIED organised and facilitated an independent farmer-led assessment of the work of Malian national agricultural research programmes on plant breeding and seed management, and of an international centre for agricultural research member of CGIAR (ICRISAT). As shown in table hereafter, recommendations expressed by farmers –both men and women– in agricultural research and extension services contrast very much with the mainstream practices of influential actors promoting the sustainable intensification of agriculture such as USAID, FAO, CGIAR, the Gates Foundations and others.

Table: Comparison between recommendations by West African small farmers' citizen juries and the practices of organizations promoting sustainable intensification.

Citizen juries composed of West African small farmers and processors	Organizations promoting sustainable intensification and allied concepts
Involve farmers in every stage of creating and selecting crop varieties.	Strategic direction for creating crop varieties set by scientists, industry and funders.
Involve producers, users and consumers (both women and men) in controlling, designing, conducting and monitoring research activities.	Mainly involve scientists, experts and funders in controlling, designing and monitoring research.
Focus on improving the productivity of local varieties, e.g. through growing practices, land use and soil fertility management.	Focus on developing new crop varieties.
Promote the use, exchange, and storage of local seeds. Avoid hybrid seeds and genetically modified organisms.	Promote improved varieties, hybrid seeds and genetically modified organisms.
Use natural mineral resources and compost; integrated pest management; and mixed cropping.	Some agencies are promoting this approach (FAO, some CGIAR projects). Others are encouraging use of artificial fertilisers and pesticides (e.g. Feed the Future, New Vision for Agriculture, some conservation agriculture projects).

Develop mechanisms to help protect the local market and local produce from unfair competition from imported products.	Increase involvement of small farmers in global supply chains and markets (New Vision for Agriculture; USAID; Gates Foundation).
Build on and disseminate farmers' agro-ecological knowledge and innovations.	(FAO; some CGIAR projects; New Vision for Agriculture; USAID). Some projects do use participatory approaches to build on farmer knowledge.

Current trends to shape mainstream agricultural research and extension services in favor of industrial farming result from a movement consisting in their 'privatization', understood as their reconfiguration or redeployment in the service of actors who benefit the most from the agro-industrial model, such as the pesticides and transgenic industries. This movement builds on the disproportionate lobby power enjoyed by those actors for influencing key policy makers in comparison to the much weaker influencing capacity of supporters of agroecology. As illustrated by a recent analysis by the Alliance for Democratising Agricultural Research in South Asia (ADARSA) and IIED in the context of South Asia, it proceeds from different levers including the rise of the private sector Research and Development (R&D), the general decline in public research funds for agriculture, the pressure for public institutions to generate income, the advent of the IP system, or the commodification of genetic resources. Foreign as well as domestic corporate players in the private sector have become important actors in R&D both through their own research, and by penetrating the public agricultural R&D sector in various ways. The 'privatization' of agricultural research and extension services can take the form of Private Public Partnerships (PPPs), whose existing examples in the realm of biological and agricultural sciences include the alliance between Novartis and the University of California to support basic agricultural genomic research or a partnership between Monsanto Inc. and the Indian government for developing hybrid basmati rice.

Corporate private actors benefitting from industrial agriculture focus on a limited range of (standardized) technologies that are profitable to them; clearly, agroecological approaches are not included since they are knowledge-intensive and need to be adapted to local conditions. Those actors cannot have any objective interest in contributing to scale-up agricultural approaches whose further expansion would necessarily reduce significantly their economic profitability. It is hard indeed to imagine how companies selling synthetic pesticides or transgenic crops, for example, could find any economic benefit in a substantial reduction of their use worldwide. This explains why genetic engineering has benefitted much more from PPPs in the agricultural sector than agroecology, since PPPs have been launched on technological trajectories in which private firms had an interest.

Current tendencies towards the 'privatization' of agricultural research and extension services must be reversed. As the guardian of the general public interest, it is time for public authorities to reinvest and prioritize research and extension services in agroecological approaches, first and foremost because of the considerable and largely untapped potential they offer.

Taking Specific Actions for Empowering Women

Rural women face numerous obstacles and gender inequities affecting their daily life, including their lack of recognition as productive farmers, unequal access to land, water, credit and other productive resources, poor access to training, extension services, or benefits from new agricultural research and technologies all of which underestimate the range of farming tasks for which they are responsible and are often incompatible with their specific needs. As noted by the Rapporteur Special on the right to food, Olivier De Schutter, gender issues are incorporated in less than 10 % of development assistance in agriculture, and women farmers receive only 5 % of extension services worldwide.

From an agroecological perspective, given their crucial role in seeds and biodiversity management, and as custodians of traditional knowledge, women's contribution is essential for successfully preserving natural resources and adapting agriculture to climate change, and they should be recognized as the innovation leaders for achieving sustainability. We have also seen that agroecology has a great potential for empowering women, but that realizing this potential is not automatic. This will require targeted actions specifically designed for tackling all the various gender abuses they are facing.

Improving Agricultural and Food Governance

Beyond all the measures described above, scaling-up agroecology also implies addressing the crucial challenge of agriculture and food governance. The challenge is essentially two-fold:

1) Improving policy coherence and,

2) Democratizing agricultural decision-making processes.

Improving Policy Coherence

In its report entitled investing in smallholder agriculture for food security, prepared to inform the adoption of recommendations by the CFS on investing in smallholder agriculture at its 40th Session in October 2013, the HLPE stressed the need to increase coherence among policies and ministries in charge of various matters impacting peasants. Increased coherence means essentially that the different policies concerned should support rather than hinder each other. For example, "investments in appropriate research and extension will not necessarily lead to improvements unless investments are also made in accessing and creating new appropriate markets. Similarly, investments in infrastructure work better if they support the models of production and markets that are appropriate to smallholders and, further, these investments would not reach their aim unless investments are also made in securing tenure rights". Furthermore, increased policy coherence is needed for taking better into account and support the multi-functionality of peasant agricultures, since "traditional ministries of agriculture are typically insufficient in fulfilling this function. Experience shows that the efficiency of specific sectoral or ministerial policies is mutually enhanced by their coordination. This often calls for specific national level governance and coordination mechanisms between different ministries, public administration and concerned stakeholders".

In order to enhance policy coherence in supportive policies for smallholders, the HLPE recommended that governments develop national Smallholder Investment Strategies: "Governments should design and implement medium- and long-term strategies, with the accompanying set of policies and budgets, to increase the capacity of the smallholder sector to fulfil its multifunctional roles in national development. These roles include contributing to growth, maintaining employment, reducing poverty, enhancing the sustainable management of natural resources and achieving food security. These National Smallholder Investment Strategies should be solidly grounded in participatory processes involving first and foremost the smallholder organizations and all concerned stakeholders". In October 2013, CFS Member States and other participants globally followed the recommendation49. The challenge is now to implement it.

Implementing it would mean putting relevant policies and regulations in line with creating an environment that allows smallholders to fulfil their multifunctional roles and develop themselves as sustainably as possible. This would imply reviewing and revising accordingly a range of various policies which negatively impact them directly or indirectly. Those policies include for example biofuels mandates and subsidies, national and regional agricultural policies, as well as bilateral and regional trade and investments agreements. Greater consistency with the requirements of smallholders' strengthening and sustainable development is an integral part of the CFS vision and roles to become "the foremost inclusive international and intergovernmental platform for a broad range of committed stakeholders to work together in a coordinated manner and in support of countryled processes towards the elimination of hunger and ensuring food security and nutrition for all human beings".

Democratizing Decision-making Processes

In the long term, all the above mentioned recommendations are essential for successfully scaling-up agroecological approaches at a higher stage. There is yet another even more important challenge: democratizing relevant decision making processes in all areas that contribute to shape the dynamics of agricultural and food systems. And for good reason: ultimately, the majority of obstacles to the diffusion of agroecological approaches are a result of a democratic deficit in relevant decisions making bodies. Throughout the world, to various degrees depending on the scale and region considered, decisions that shape agricultural and food systems are indeed disproportionately influenced by vested interests of a minority of actors to the detriment of the general public interest, of sustainable development and of the fundamental rights of a majority of populations. These actors are the proponents as well as the beneficiaries of the current corporate agro-industrial food system. They include 'traditional' actors of the agri-food value chains such as global food retailers, food processors, commodity traders, the pesticide or transgenic industries. They also include other actors, not active or traditionally active in food and agriculture, such as pension funds, companies in the automotive industry, or oil companies, who can exert a strong, indirect influence on various policies that directly or indirectly shape the dynamic of the agri-food system, be they agricultural, energy, trade, or financial policies. These actors invest considerable means to protect their particular interests against any decision that may threaten them. Their influence over decision-making processes seriously harms the global sustainability of the food and agricultural system: its capacity to feed the world, preserve biodiversity, tackle climate change, eradicate poverty and address other sustainability challenges. Below provides a few examples of the corporate influence over the food and agricultural system.

The large influence of transnational companies (TNCs) that make-up the world seed, agrochemical and biotechnology markets, which have a vested interest in maintaining a monoculture-focused, carbon-intensive industrial approach to agriculture dependent on external inputs, is obvious. Powerful commercial interests of agribusinesses especially hamper the scaling-up of agroecological transitions to meet sustainability challenges.

In South Asia, a second Green Revolution, along with a 'Gene Revolution', is being rolled out jointly by government and the large corporate sector. In 2012, the national Indian Ministry of Agriculture, for example, has made clear its intention "to fully extend green revolution to all the low productivity areas of eastern region where there is good potential to harness ample natural resources in order to achieve food security and agricultural sustainability". By contrast, small-scale farmers, who are both the primary practitioners of agroecology and the first beneficiaries of its expanded use, are most often systematically marginalized in policy decisions.

Corporate Influence over the Food and Agricultural System

Corporate actors strongly influence the directions of various national and international policies that contribute to shape the dynamic of the agricultural and food system. Examples include:

- A former corporate counsel for the pesticides and biotechnology company, Dupont, was appointed in January 2011 to serve as general counsel for the USDA. Soon after, the USDA proposed a dramatic reduction in agency responsibility for regulating genetic engineered (GE) crops;

- For all of 2007, agribusiness giants Syngenta and Monsanto spent $1.2 million and $4.5 million respectively on lobbying the US federal government on pesticide legislation, biofuels, patent laws and other issues, according to regulatory filings;

- Monsanto and its affiliates lobbied Indonesian legislators in the 1990s to support GE crops. In 2005, the firm was fined $1,5 million by the United States Department of Justice for violating the Foreign Corrupt Practices Act by bribing a senior Indonesian Environment Ministry official;

- The first version of the Agreement on Agriculture of the WTO was drafted by Dan Amstutz, then a director of Cargill and president of the North-American association of grain exporters, before becoming under-secretary of the US Department of Agriculture charged with market support programs, then chief agricultural negotiator during the Uruguay Round, then president of Amstutz & Company, a consultancy firm specialized in agri-business and international trade and finally president of the board of a common enterprise of ADM, Cargill, Cenex Harvest States, DuPont and Louis Dreyfuss;

- The world's largest investment banks, trading companies active in the agrifood sector, and other companies speculating on derivative markets invested enormous efforts, first to prevent, then to water down the reform of the U.S. financial regulatory system (Dodd-Frank Wall Street Reform and Consumer Protection Act) signed by President Barack Obama on the 21st of July 2010 . For the period prior to the adoption of the Dodd-Frank Act, Wall Street employed 2,000 lobbyists in Washington with a 600 $million budget. Since its adoption, they have successfully weakened its implementation;

- The European biofuels industry has achieved a major victory with the adoption in 2009 of the European Directive on renewable energy, which includes a mandatory target of 10% renewable energy in the EU transport sector for 2020 (in practice essentially reached through biofuels). This policy offers bright prospects for growth for concerned companies, including companies active in the agrifood, biochemical, oil and automotive sectors;

- Following controversy over its close ties with industry, the European Food Safety Authority (EFSA), which is "committed to ensuring that Europe's food is safe", has implemented in 2013 a new policy designed to ensure the independence of its scientific panels. Experts involved in these panels play a crucial role in decisions key to the health and safety of Europe's food supply chain. Yet, according to a recent study of the Corporate Europe Observatory (CEO), serious conflicts of interest remain: over half of the 209 scientists sitting on the agency's panels have direct or indirect ties with the industries they are meant to regulate;

- In Bangladesh, the state has been unable at times to supply seeds to farmers because of unpaid dues to contracted seed growers. Yet, the state still does not encourage farm-saved seeds (FSS). On the contrary, influenced by interests of the industry, new laws and policies are creating an environment for the private seed companies to sell their seeds, while there is no regulatory framework for developing and expanding local seed systems for crops or varieties important to small-scale farmers.

The disproportionate capacity of a minority of actors to shape agricultural and food systems relies on their huge 'market power'. Applied to food and agriculture, the 'market power' of a given actor could be broadly defined as its capacity to impose its practices and requirements on others, in particular in the framework of agrifood commercial transactions along the value chains (in terms of buying or selling prices, delivery times, production standards), and/or to influence to its advantage the policies and laws that potentially impact its objective interests and contribute to shape directly or indirectly the agricultural and food system. Market power of a given actor depends mainly on its size, the concentration (horizontal and vertical) of its sector of activity, its financial capital, and its social capital (extent to which the actor nurtures close social/cultural ties with policy makers).

Democratizing decision-making processes and in particular increasing the active participation of peasants in decisions that affect them and shape agricultural and food systems should be an absolute priority for scaling-up agroecological approaches at a higher stage. This is a key stepping-stone for truly overcoming obstacles for more sustainable agricultural and food systems. Active participation of peasants, and especially women, must be ensured at local, regional (subnational), national and international levels. Real participation is crucial to ensure that all relevant policies are truly responsive of the needs of vulnerable groups and for empowering them. Public authorities have the obligation to take strong actions for dismantling the disproportionate market power of those using their influence to highjack and format agricultural and food systems to serve their own private interests.

Principles and Strategies of Agroecology

The concept of sustainable agriculture is a relatively recent response to the decline in the quality of the natural resource base associated with modern agriculture. Today, the question of agricultural production has evolved from a purely technical one to a more complex one characterized by social, cultural, political and economic dimensions. The concept of sustainability although controversial and diffuse due to existing conflicting definitions and interpretations of its meaning, is useful because it captures a set of concerns about agriculture which is conceived as the result of the coevolution of socioeconomic and natural systems. A wider understanding of the agricultural context requires the study between agriculture, the global environment and social systems given that agricultural development results from the complex interaction of a multitude of factors. It is through this deeper understanding of the ecology of agricultural systems that doors will open to new management options more in tune with the objectives of a truly sustainable agriculture.

The sustainability concept has prompted much discussion and has promoted the need to propose major adjustments in conventional agriculture to make it more environmentally, socially and economically viable and compatible. Several possible solutions to the environmental problems created by capital and technology intensive farming systems have been proposed and research is currently in progress to evaluate alternative systems. The main focus lies on the reduction or elimination of agrochemical inputs through changes in management to assure adequate plant nutrition and plant protection through organic nutrient sources and integrated pest management, respectively.

Although hundreds of more environmentally prone research projects and technological development attempts have taken place, and many lessons have been learned, the thrust is still highly technological, emphasizing the suppression of limiting factors or the symptoms that mask an ill producing agroecosystem. The prevalent philosophy is that pests, nutrient deficiencies or other factors are the cause of low productivity, as opposed to the view that pests or nutrients only become limiting if conditions in the agroecosystem are not in equilibrium. For this reason, there still prevails a narrow view that specific causes affect productivity, and overcoming the limiting factor via new technologies, continues to be the main goal. This view has diverted agriculturists from realizing that limiting factors only represent symptoms of a more systemic disease inherent to unbalances within the agroecosystem and from an appreciation of the context and complexity of agroecological processes thus underestimating the root causes of agricultural limitations.

On the other hand, the science of agroecology, which is defined as the application of ecological concepts and principles to the design and management of sustainable agroecosystems, provides a framework to assess the complexity of agroecosystems. The idea of agroecology is to go beyond the use of alternative practices and to develop agroecosystems with the minimal dependence on high agrochemical and energy inputs, emphasizing complex agricultural systems in which ecological interactions and synergisms between biological components provide the mechanisms for the systems to sponsor their own soil fertility, productivity and crop protection.

Principles of Agroecology

In the search to reinstate more ecological rationale into agricultural production, scientists and developers have disregarded a key point in the development of a more self-sufficient and sustaining agriculture: a deep understanding of the nature of agroecosystems and the principles by which they function. Given this limitation, agroecology has emerged as the discipline that provides the basic ecological principles for how to study, design and manage agroecosystems that are both productive and natural resource conserving, and that are also culturally sensitive, socially just and economically viable.

Agroecology goes beyond a one-dimensional view of agroecosystems - their genetics, agronomy, edaphology, and so on, - to embrace an understanding of ecological and social levels of co-evolution, structure and function. Instead of focusing on one particular component of the agroecosystem, agroecology emphasizes the interrelatedness of all agroecosystem components and the complex dynamics of ecological processes.

Agroecosystems are communities of plants and animals interacting with their physical and chemical environments that have been modified by people to produce food, fibre, fuel and other products for human consumption and processing. Agroecology is the holitstic study of agroecosystems, including all environmental and human elements. It focuses on the form, dynamics and functions of their interrelationships and the processes in which they are involved. An area used for agricultural production, e.g. a field, is seen as a complex system in which ecological processes found under natural conditions also occur, e.g. nutrient cycling, predator/prey interactions, competition, symbiosis and successional changes. Implicit in agroecological research is the idea that, by understanding these ecological relationships and processes, agroecosystems can be manipulated to improve production and to produce more sustainably, with fewer negative environmental or social impacts and fewer external inputs.

The design of such systems is based on the application of the following ecological principles:

1. Enhance recycling of biomass and optimizing nutrient availability and balancing nutrient flow.

2. Securing favorable soil conditions for plant growth, particularly by managing organic matter and enhancing soil biotic activity.

3. Minimizing losses due to flows of solar radiation, air and water by way of microclimate management, water harvesting and soil management through increased soil cover.

4. Species and genetic diversification of the agroecosystem in time and space.

5. Enhance beneficial biological interactions and synergisms among agrobiodiversity components thus resulting in the promotion of key ecological processes and services.

These principles can be applied by way of various techniques and strategies. Each of these will have different effects on productivity, stability and resiliency within the farm system, depending on the local opportunities, resource constraints and, in most cases, on the market. The ultimate goal of agroecological design is to integrate components so that overall biological efficiency is improved, biodiversity is preserved, and the agroecosystem productivity and its self-sustaining capacity is maintained. The goal is to design a quilt of agroecosystems within a landscape unit, each mimicking the structure and function of natural ecosystems.

Biodiversification of Agroecosystems

From a management perspective, the agroecological objective is to provide balanced environments, sustained yields, biologically mediated soil fertility and natural pest regulation through the design of diversified agroecosystems and the use of low-input technologies (Gleissman 1998). Agroecologists are now recognizing that intercropping, agroforestry and other diversification methods mimic natural ecological processes, and that the sustainability of complex agroecosystems lies in the ecological models they follow. By designing farming systems that mimic nature, optimal use can be made of sunlight, soil nutrients and rainfall.

Agroecological management must lead management to optimal recycling of nutrients and organic matter turnover, closed energy flows, water and soil conservation and balance pest-natural enemy populations. The strategy exploits the complementarities and synergisms that result from the various combinations of crops, tree and animals in spatial and temporal arrangements.

In essence, the optimal behavior of agroecosystems depends on the level of interactions between the various biotic and abiotic components. By assembling a functional biodiversity it is possible to initiate synergisms which subsidize agroecosystem processes by providing ecological services such as the activation of soil biology, the recycling of nutrients, the enhancement of beneficial arthropods and antagonists, and so on. Today there is a diverse selection of practices and technologies available, and which vary in effectiveness as well as in strategic value. Key practices are those of a preventative nature and which act by reinforcing the "immunity" of the agroecosystem through a series of mechanisms.

Various strategies to restore agricultural diversity in time and space include crop rotations, cover crops, intercropping, crop/livestock mixtures, and so on, which exhibit the following ecological features:

1. Crop Rotations: Temporal diversity incorporated into cropping systems, providing crop nutrients and breaking the life cycles of several insect pests, diseases, and weed life cycles.

2. Polycultures: Complex cropping systems in which two or more crop species are planted within sufficient spatial proximity to result in competition or complementation, thus enhancing yields.

3. Agroforestry Systems: An agricultural system where trees are grown together with annual crops and/or animals, resulting in enhanced complementary relations between components increasing multiple use of the agroecosystem.

4. Cover Crops: The use of pure or mixed stands of legumes or other annual plant species under fruit trees for the purpose of improving soil fertility, enhancing biological control of pests, and modifying the orchard microclimate.

5. Animal integration in agroecosystems aids in achieving high biomass output and optimal recycling.

All of the above diversified forms of agroecosystems share in common the following features:

- Maintain vegetative cover as an effective soil and water conserving measure, met through the use of no-till practices, mulch farming, and use of cover crops and other appropriate methods.

- Provide a regular supply of organic matter through the addition of organic matter (manure, compost, and promotion of soil biotic activity).

- Enhance nutrient recycling mechanisms through the use of livestock systems based on legumes, etc.

- Promote pest regulation through enhanced activity of biological control agents achieved by introducing and/or conserving natural enemies and antagonists.

Research on diversified cropping systems underscores the great importance of diversity in an agricultural setting. Diversity is of value in agroecosystems for a variety of reasons:

- As diversity increases, so do opportunities for coexistence and beneficial interactions between species that can enhance agroecosystem sustainability.

- Greater diversity often allows better resource-use efficiency in an agroecosystem. There is better system-level adaptation to habitat heterogeneity, leading to complementarity in crop species needs, diversification of niches, overlap of species niches, and partitioning of resources.

- Ecosystems in which plant species are intermingled possess an associated resistance to herbivores as in diverse systems there is a greater abundance and diversity of natural enemies of pest insects keeping in check the populations of individual herbivore species.

- A diverse crop assemblage can create a diversity of microclimates within the cropping system that can be occupied by a range of non-crop organisms - including beneficial predators, parasites, pollinators, soil fauna and antagonists - that are of importance for the entire system.

- Diversity in the agricultural landscape can contribute to the conservation of biodiversity in surrounding natural ecosystems.

- Diversity in the soil performs a variety of ecological services such as nutrient recycling and detoxification of noxious chemicals and regulation of plant growth.

- Diversity reduces risk for farmers, especially in marginal areas with more unpredictable environmental conditions. If one crop does not do well, income from others can compensate.

Agroecology and the Design of Sustainable Agroecosystems

Most people involved in the promotion of sustainable agriculture aim at creating a form of agriculture that maintains productivity in the long term by:

- Optimizing the use of locally available resources by combining the different components of the farm system, i.e. plants, animals, soil, water, climate and people, so that they complement each other and have the greatest possible synergetic effects;

- Reducing the use of off-farm, external and non-renewable inputs with the greatest potential to damage the environment or harm the health of farmers and consumers, and a more targeted use of the remaining inputs used with a view to minimizing variable costs;

- Relying mainly on resources within the agroecosystem by replacing external inputs with nutrient cycling, better conservation, and an expanded use of local resources;

- Improving the match between cropping patterns and the productive potential and environmental constraints of climate and landscape to ensure long-term sustainability of current production levels;

- Working to value and conserve biological diversity, both in the wild and in domesticated landscapes, and making optimal use of the biological and genetic potential of plant and animal species; and

- Taking full advantage of local knowledge and practices, including innovative approaches not yet fully understood by scientists although widely adopted by farmers.

Agroecology provides the knowledge and methodology necessary for developing an agriculture that is on the one hand environmentally sound and on the other hand highly productive, socially equitable and economically viable. Through the application of agroecological principles, the basic challenge for sustainable agriculture to make better use of internal resources can be easily achieved by minimizing the external inputs used, and preferably by regenerating internal resources more effectively through diversification strategies that enhance synergisms among key components of the agroecosystem.

The ultimate goal of agroecological design is to integrate components so that overall biological efficiency is improved, biodiversity is preserved, and the agroecosystem productivity and its self-regulating capacity are maintained. The goal is to design an agroecosystem that mimics the structure and function of local natural ecosystems; that is, a system with high species diversity and a biologically active soil, one that promotes natural pest control, nutrient recycling and high soil cover to prevent resource losses.

Elements of Agroecology

As an analytical tool, the 10 Elements can help countries to operationalise agroecology. By identifying important properties of agroecological systems and approaches, as well as key considerations in developing an enabling environment for agroecology, the 10 Elements are a guide for policymakers, practitioners and stakeholders in planning, managing and evaluating agroecological transitions.

Agroecological systems are highly diverse. From a biological perspective, agroecological systems optimize the diversity of species and genetic resources in different ways. For example, agroforestry systems organize crops, shrubs, and trees of different heights and shapes at different levels or strata, increasing vertical diversity.

Intercropping combines complementary species to increase spatial diversity. Crop rotations, often including legumes, increase temporal diversity. Crop–livestock systems rely on the diversity of local breeds adapted to specific environments. In the aquatic world, traditional fish polyculture farming, Integrated Multi-Trophic Aquaculture (IMTA) or rotational crop-fish systems follow the same principles to maximising diversity.

Increasing biodiversity contributes to a range of production, socio-economic, nutrition and environmental benefits. By planning and managing diversity, agroecological approaches enhance the provisioning of ecosystem services, including pollination and soil health, upon which agricultural production depends. Diversification can increase productivity and resource-use efficiency by optimizing biomass and water harvesting.

Agroecological diversification also strengthens ecological and socio-economic resilience, including by creating new market opportunities. For example, crop and animal diversity reduces the risk of failure in the face of climate change.

Mixed grazing by different species of ruminants reduces health risks from parasitism, while diverse local species or breeds have greater abilities to survive, produce and maintain reproduction levels in harsh environments. In turn, having a variety of income sources from differentiated and new markets, including diverse products, local food processing and agritourism, helps to stabilize household incomes.

Consuming a diverse range of cereals, pulses, fruits, vegetables, and animal-source products contributes to improved nutritional outcomes. Moreover, the genetic diversity of different varieties, breeds and species is important in contributing macronutrients, micronutrients and other bioactive compounds to human diets. For example, in Micronesia, reintroducing an underutilized traditional variety of orange-fleshed banana with 50 times more beta-carotene than the widely available commercial white-fleshed banana proved instrumental in improving health and nutrition.

At the global level, three cereal crops provide close to 50 percent of all calories consumed, while the genetic diversity of crops, livestock, aquatic animals and trees continues to be rapidly lost.

Agroecology can help reverse these trends by managing and conserving agro-biodiversity, and responding to the increasing demand for a diversity of products that are eco-friendly. One such example is 'fish-friendly' rice produced from irrigated, rainfed and deepwater rice ecosystems, which values the diversity of aquatic species and their importance for rural livelihoods.

Agroecology depends on context-specific knowledge. It does not offer fixed prescriptions – rather, agroecological practices are tailored to fit the environmental, social, economic, cultural and political context. The co-creation and sharing of knowledge plays a central role in the process of developing and implementing agroecological innovations to address challenges across food systems including adaptation to climate change.

Through the co-creation process, agroecology blends traditional and indigenous knowledge, producers' and traders' practical knowledge, and global scientific knowledge.

Producer's knowledge of agricultural biodiversity and management experience for specific contexts as well as their knowledge related to markets and institutions are absolutely central in this process.

Education – both formal and non-formal – plays a fundamental role in sharing agroecological innovations resulting from co-creation processes. For example, for more than 30 years, the horizontal campesino a campesino movement has played a pivotal role in sharing agroecological knowledge, connecting hundreds of thousands of producers in Latin America.12 In contrast, top-down models of technology transfer have had limited success.

Promoting participatory processes and institutional innovations that build mutual trust enables the co-creation and sharing of knowledge, contributing to relevant and inclusive agroecology transition processes.

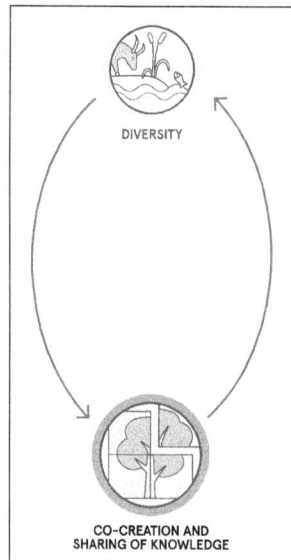

DIVERSITY

CO-CREATION AND
SHARING OF KNOWLEDGE

Agroecology pays careful attention to the design of diversified systems that selectively combine annual and perennial crops, livestock and aquatic animals, trees, soils, water and other components on farms and agricultural landscapes to enhance synergies in the context of an increasingly changing climate.

Building synergies in food systems delivers multiple benefits. By optimizing biological synergies, agroecological practices enhance ecological functions, leading to greater resource-use efficiency and resilience. For example, globally, biological nitrogen fixation by pulses in intercropping systems or rotations generates close to USD 10 million savings in nitrogen fertilizers every year, while contributing to soil health, climate change mitigation and adaptation. Furthermore, about 15 percent of the nitrogen applied to crops comes from livestock manure, highlighting synergies resulting from crop–livestock integration. In Asia, integrated rice systems combine rice cultivation with the generation of other products such as fish, ducks and trees. By maximising synergies, integrated rice systems significantly improve yields, dietary diversity, weed control, soil structure and fertility, as well as providing biodiversity habitat and pest control.

At the landscape level, synchronization of productive activities in time and space is necessary to enhance synergies. Soil erosion control using Calliandra hedgerows is common in integrated agroecological systems in the East African Highlands. In this example, the management practice of periodic pruning reduces tree competition with crops grown between hedgerows and at the same time provides feed for animals, creating synergies between the different components. Pastoralism and extensive livestock grazing systems manage complex interactions between people, multi-species herds and variable environmental conditions, building resilience and contributing to ecosystem services such as seed dispersal, habitat preservation and soil fertility.

While agroecological approaches strive to maximise synergies, trade-offs also occur in natural and human systems. For example, the allocation of resource use or access rights often involve trade-offs. To promote synergies within the wider food system, and best manage trade-offs, agroecology emphasizes the importance of partnerships, cooperation and responsible governance, involving different actors at multiple scales.

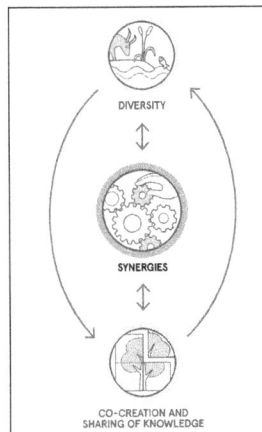

Increased resource-use efficiency is an emergent property of agroecological systems that carefully plan and manage diversity to create synergies between different system components. For example, a key efficiency challenge is that less than 50 percent of nitrogen fertilizer added globally to cropland is converted into harvested products and the rest is lost to the environment causing major environmental problems.

Agroecological systems improve the use of natural resources, especially those that are abundant and free, such as solar radiation, atmospheric carbon and nitrogen.

By enhancing biological processes and recycling biomass, nutrients and water, producers are able to use fewer external resources, reducing costs and the negative environmental impacts of their use.

Ultimately, reducing dependency on external resources empowers producers by increasing their autonomy and resilience to natural or economic shocks.

One way to measure the efficiency of integrated systems is by using Land Equivalent Ratios (LER). LER compares the yields from growing two or more components (e.g. crops, trees, animals) together with yields from growing the same components individually. Integrated agroecological systems frequently demonstrate higher LERs.

Agroecology thus promotes agricultural systems with the necessary biological, socio-economic and institutional diversity and alignment in time and space to support greater efficiency.

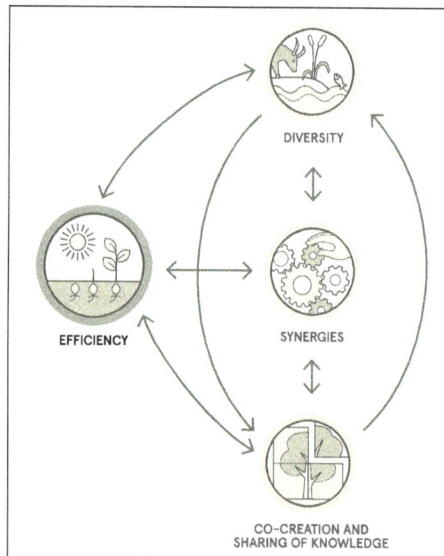

Waste is a human concept – it does not exist in natural ecosystems. By imitating natural ecosystems, agroecological practices support biological processes that drive the recycling of nutrients, biomass and water within production systems, thereby increasing resourceuse efficiency and minimizing waste and pollution.

Recycling can take place at both farm-scale and within landscapes, through diversification and building of synergies between different components and activities. For example, agroforestry systems that include deep rooting trees can capture nutrients lost beyond the roots of annual crops. Crop–livestock systems promote recycling of organic materials by using manure for composting or directly as fertilizer, and crop residues and by-products as livestock feed.

Nutrient cycling accounts for 51 percent of the economic value of all non-provisioning ecosystem services, and integrating livestock plays a large role in this. Similarly, in rice–fish systems, aquatic animals help to fertilize the rice crop and reduce pests, reducing the need for external fertilizer or pesticide inputs.

Recycling delivers multiple benefits by closing nutrient cycles and reducing waste that translates into lower dependency on external resources, increasing the autonomy of producers and reducing their vulnerability to market and climate shocks. Recycling organic materials and by-products offers great potential for agroecological innovations.

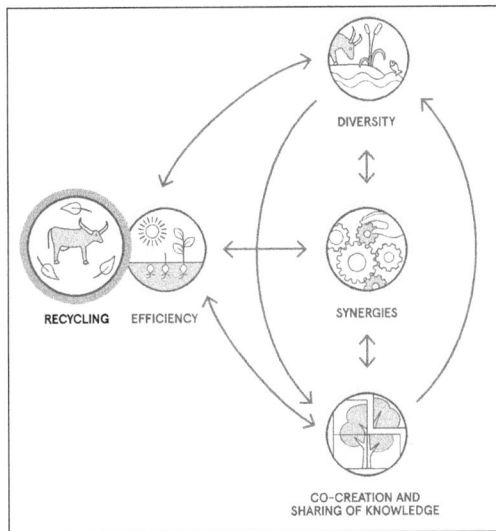

DIVERSITY

RECYCLING   EFFICIENCY

SYNERGIES

CO-CREATION AND
SHARING OF KNOWLEDGE

Diversified agroecological systems are more resilient – they have a greater capacity to recover from disturbances including extreme weather events such as drought, floods or hurricanes, and to resist pest and disease attack.

Following Hurricane Mitch in Central America in 1998, biodiverse farms including agroforestry, contour farming and cover cropping retained 20–40 percent more topsoil, suffered less erosion and experienced lower economic losses than neighbouring farms practicing conventional monocultures.

By maintaining a functional balance, agroecological systems are better able to resist pest and disease attack. Agroecological practices recover the biological complexity of agricultural systems and promote the necessary community of interacting organisms to self regulate pest outbreaks.

On a landscape scale, diversified agricultural landscapes have a greater potential to contribute to pest and disease control functions.

Agroecological approaches can equally enhance socio-economic resilience. Through diversification and integration, producers reduce their vulnerability should a single crop, livestock species or other commodity fail.

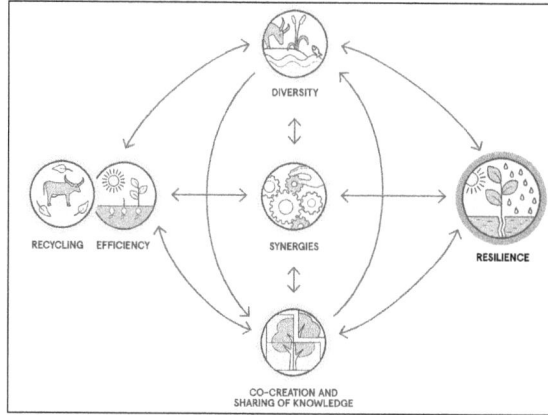

By reducing dependence on external inputs, agroecology can reduce producers' vulnerability to economic risk. Enhancing ecological and socioeconomic resilience goes hand-in-hand – after all, humans are an integral part of ecosystems.

Agroecology places a strong emphasis on human and social values, such as dignity, equity, inclusion and justice all contributing to the improved livelihoods dimension of the SDGs. It puts the aspirations and needs of those who produce, distribute and consume food at the heart of food systems. By building autonomy and adaptive capacities to manage their agro-ecosystems, agroecological approaches empower people and communities to overcome poverty, hunger and malnutrition, while promoting human rights, such as the right to food, and stewardship of the environment so that future generations can also live in prosperity.

Agroecology seeks to address gender inequalities by creating opportunities for women. Globally, women make up almost half of the agricultural workforce. They also play a vital role in household food security, dietary diversity and health, as well as in the conservation and sustainable use of biological diversity. In spite of this, women remain economically marginalised and vulnerable to violations of their rights, while their contributions often remain unrecognized.

Agroecology can help rural women in family farming agriculture to develop higher levels of autonomy by building knowledge, through collective action and creating opportunities for commercialization. Agroecology can open spaces for women to become more autonomous and empower them at household, community levels and beyond – for instance, through participation in producer groups. Women's participation is essential for agroecology and women are frequently the leaders of agroecology projects.

In many places around the world, rural youth face a crisis of employment. Agroecology provides a promising solution as a source of decent jobs. Agroecology is based on a different way of agricultural production that is knowledge intensive, environmentally friendly, socially responsible, innovative, and which depends on skilled labour. Meanwhile, rural youth around the world possess energy, creativity and a desire to positively change their world. What they need is support and opportunities.

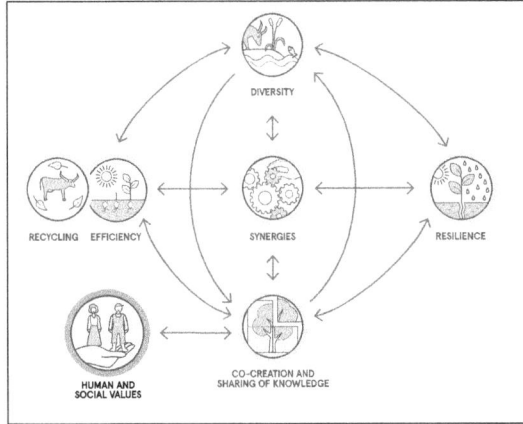

As a bottom-up, grassroots paradigm for sustainable rural development, agroecology empowers people to become their own agents of change.

Agriculture and food are core components of human heritage. Hence, culture and food traditions play a central role in society and in shaping human behaviour. However, in many instances, our current food systems have created a disconnection between food habits and culture. This disconnection has contributed to a situation where hunger and obesity exist side by side, in a world that produces enough food to feed its entire population.

Almost 800 million people worldwide are chronically hungry and 2 billion suffer micronutrient deficiencies. Meanwhile, there has been a rampant rise in obesity and diet-related diseases; 1.9 billion people are overweight or obese and non-communicable diseases (cancer, cardiovascular disease, diabetes) are the number one cause of global mortality.

To address the imbalances in our food systems and move towards a zero hunger world, increasing production alone is not sufficient.

Agroecology plays an important role in re-balancing tradition and modern food habits, bringing them together in a harmonious way that promotes healthy food production and consumption, supporting the right to adequate food. In this way, agroecology seeks to cultivate a healthy relationship between people and food.

Cultural identity and sense of place are often closely tied to landscapes and food systems. As people and ecosystems have evolved together, cultural practices and indigenous and

traditional knowledge offer a wealth of experience that can inspire agroecological solutions. For example, India is home to an estimated 50 000 indigenous varieties of rice –
bred over centuries for their specific taste, nutrition and pestresistance properties, and
their adaptability to a range of conditions. Culinary traditions are built around these
different varieties, making use of their different properties. Taking this accumulated
body of traditional knowledge as a guide, agroecology can help realise the potential of
territories to sustain their peoples.

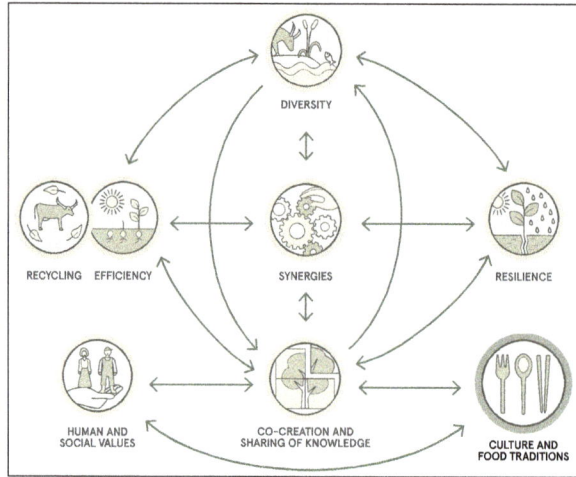

Agroecology calls for responsible and effective governance to support the transition to
sustainable food and agricultural systems. Transparent, accountable and inclusive governance mechanisms are necessary to create an enabling environment that supports
producers to transform their systems following agroecological concepts and practices. Successful examples include school feeding and public procurement programmes,
market regulations allowing for branding of differentiated agroecological produce, and
subsidies and incentives for ecosystem services.

Land and natural resources governance is a prime example. The majority of the world's
rural poor and vulnerable populations heavily rely on terrestrial and aquatic biodiversity and ecosystem services for their livelihoods, yet lack secure access to these resources.

Agroecology depends on equitable access to land and natural resources – a key to social
justice, but also in providing incentives for the long-term investments that are necessary to protect soil, biodiversity and ecosystem services.

Agroecology is best supported by responsible governance mechanisms at different
scales. Many countries have already developed national level legislation, policies and
programmes that reward agricultural management that enhances biodiversity and the
provision of ecosystem services. Territorial, landscape and community level governance, such as traditional and customary governance models, is also extremely important to foster cooperation between stakeholders, maximising synergies while reducing
or managing trade-offs.

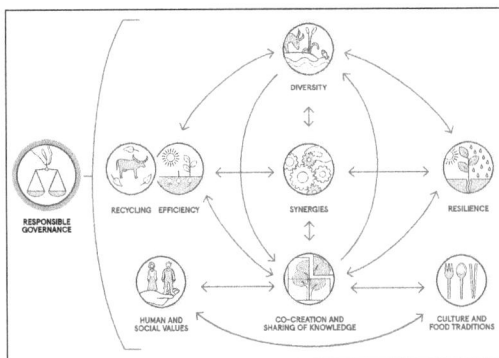

Agroecology seeks to reconnect producers and consumers through a circular and solidarity economy that prioritizes local markets and supports local economic development by creating virtuous cycles.

Agroecological approaches promote fair solutions based on local needs, resources and capacities, creating more equitable and sustainable markets. Strengthening short food circuits can increase the incomes of food producers while maintaining a fair price for consumers. These include new innovative markets, alongside more traditional territorial markets, where most smallholders market their products.

Social and institutional innovations play a key role in encouraging agroecological production and consumption. Examples of innovations that help link producers and consumers include participatory guarantee schemes, local producer's markets, denomination of origin labelling, community supported agriculture and e-commerce schemes. These innovative markets respond to a growing demand from consumers for healthier diets.

Re-designing food systems based on the principles of circular economy can help address the global food waste challenge by making food value chains shorter and more resource-efficient. Currently, one third of all food produced is lost or wasted, failing to contribute to food security and nutrition, while exacerbating pressure on natural resources.

The energy used to produce food that is lost or wasted is approximately 10 percent of the world's total energy consumption, while the food waste footprint is equivalent to 3.5 Gt CO_2 of greenhouse gas emissions per year.

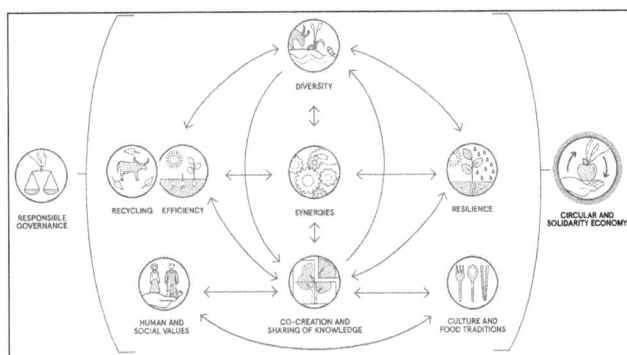

Agroecological Matrix

Among the myriad complications involved in the current food crisis, the relationship between agriculture and the rest of nature is one of the most important yet remains only incompletely analyzed. Particularly in tropical areas, agriculture is frequently seen as the antithesis of the natural world, where the problem is framed as one of minimizing land devoted to agriculture so as to devote more to conservation of biodiversity and other ecosystem services. In particular, the "forest transition model" projects an overly optimistic vision of a future where increased agricultural intensification (to produce more per hectare) and/or increased rural-to-urban migration (to reduce the rural population that cuts forest for agriculture) suggests a near future of much tropical aforestation and higher agricultural production.

The Forest Transition Model

The European colonization of eastern North America began with massive deforestation that accompanied the expansion of agriculture. But then, through industrialization and the urbanization that accompanied it, agriculture declined and forests returned. The dynamics that drove this process are evident at a broad qualitative level—wealth from agriculture drives local industrialization that, in turn, acts as a magnet for labor, which depopulates the countryside, leaving natural succession to take over. Although this general view has many complications that drive local ecological and sociopolitical dynamics, as an overview of eastern North American forest history it seems historically accurate, and has been referred to as the "forest transition model". Similar processes have been described for some European countries, the rural U.S. South, and, most importantly given its tropical location, Puerto Rico. Based on this and other examples, some have proposed that the FT model could be a framework for understanding tropical landscape dynamics in general and even be used for promoting a conservation agenda.

Although the argument is usually made in an informal qualitative sense, there is an underlying quantitative logic that drives the conclusions. Understanding that logic is helpful for understanding exactly where the argument is wrong.

Consider a defined land area of total size T divided into one portion that is agricultural (a) and another set aside for conservation (c); p represents the units of production (in energy per unit area), N_L is the local (rural) population density, and e is the energy requirements of a single person. Clearly, at equilibrium,

$$pa = N_L e, \quad or \quad a^* = N_L e / p,$$

which suggests that we can minimize a* by minimizing N_L and/or maximizing p (assuming ewill always remain constant). At its most simplistic level, this is the land-sparing argument.

The argument is elementary, based on simple accounting, suggesting that there are basically two sociopolitical-ecological forces in operation: first, a spatial concentration and intensification of agricultural production and, second, an exodus of the rural population to industrializing urban centers. Taken together, these forces reduce the demand for cropland, thus freeing marginal farmlands and leading to recovery of forests. This idea has become common and is sometimes taken as a self-evident process, worthy of paradigmatic status for conservation.

Obvious complications arise with only a slightly larger view of the population that must be serviced by agriculture. Consider, for example, that the total population, NT, consists of the sum of the rural population, NL, and the urban population (i.e., the population not involved in agricultural production but needing the products of agriculture), NU; in other words, NL + NU = NT. Modifying $pa = N_L e$, or $a^* = N_L e / p$, we have $a^* = e(NL + NU)/p$. Presuming each person works (w) land units to maintain and produce in the agricultural system, we have, at equilibrium,

$$\frac{\left(N_L + N_U\right)e}{P} = wN_L,$$

where the left-hand side is the amount of agricultural land needed to support the relevant population (NL + NU) and the right-hand side is the amount of agricultural land maintainable with NL workers working at a rate w. If the agricultural land needed is greater than the agricultural land maintainable, we see (from $\dfrac{\left(N_L + N_U\right)e}{P} = wN_L$),

$$N_U e > \left(pw - e\right)N_L,$$

and the local experience will be one of a labor shortage (because the agricultural land needed to sustain the population is greater than the available labor can sustain). Making the reasonable assumption that equilibrium will be a social goal; the FT model proposes that we can equilibrate $N_U e > \left(pw - e\right)N_L$, by increasing either w or p, which could be done with labor-saving technology or higher units of production. However, with this formulation it is evident that increasing w or p are not the only ways of equilibrating $N_U e > \left(pw - e\right)N_L$. An alternative would be to increase the local rural population (contrary to the FT model). Ironically, as rural-to-urban migration proceeds, the inequality in $N_U e > \left(pw - e\right)N_L$, becomes more accentuated and the need to increase rural population consequently increases yet further.

Consider the reverse situation, where the agricultural land needed is less than the agricultural land maintainable (i.e., the inequality in $N_U e > \left(pw - e\right)N_L$, is reversed). Here the local experience is overproduction. Equilibrating the equation can be done by decreasing w (which is easily accomplished by taking land out of production), decreasing NL, or decreasing p, the last of which is clearly contrary to the basic ideas of the FT model.

Human response to the experience of either labor shortage (relation $N_U e > (pw - e)N_L,$) or overproduction (relation $N_U e > (pw - e)N_L$, with the sign reversed) has always been complicated, with strong dependence on the way the society is organized. For example, in early barter and exchange societies where most agricultural production was for the use of the agricultural family itself, the response to overproduction is likely to be simply reducing w, that is, to take land out of production (no need to produce what you will not need). However, in more market-oriented societies, overproduction may lead to lowered market prices and the tendency by individual producers to increase production further to increase total farm revenue, or a shift to another commodity which may require more land (for example, extensive cattle pasture). In both cases, the result is the reverse of what would be expected from the simple FT model. Additionally, if production planning is keyed to current price conditions, simple nonlinearities may lead to chaotic price and production trajectories over the long haul, making it, in principle, impossible to say whether w will increase or decrease. Clearly, the social context makes an enormous difference.

Ultimately, the FT model rests on two quantitative assumptions and a seemingly logical conclusion. The two assumptions are, first, a given population density requires a certain land base to enable productive activities adequate to survival of the whole population (the "sustainable" population) and, second, the amount of food required to support that population, divided by current per-area productivity, equals the land area necessary for agricultural production (the rural population density required to support that production is the "necessary" population). The logical conclusion is that the total land area minus the area necessary for production is what is available for conservation.

The "rural-to-urban migration" part of the FT model focuses on the first assumption and notes that, with the reduction in rural population, more land will be available for conservation (fewer rural people, less use of land for agriculture, and thus natural regeneration of forest or other natural habitat). The "productivity" argument focuses on the second assumption and argues that if per-unit production could be increased, the required land base would be reduced, and consequently more land would be available for conservation (the same number of people needing food but higher productivity, thus less land for agriculture and more land for conservation). Referring again to 3, it is certainly possible for the FT model to operate, but our point is that it is not in any way quantitatively assured that it actually will. Theoretically, the issue is indeterminate. It thus makes sense to ask to what extent do real-world data suggest that recent tropical situations replay the experience of the previous examples that had given conservationists such hope (e.g., Puerto Rico or New England).

Angelsen and Kaimowitz report on detailed studies that, as might be expected from the argument presented above, sometimes support the FT model, sometimes fail to support it. Their study notes an underlying contradiction in the basic ideas of the FT model. First, "the belief that technological progress in agriculture reduces pressure on forests by allowing farmers to produce the same amount of food in a smaller area has become almost an article of faith in development and environmental circles." Second, "basic economic theory suggests that technological progress makes agriculture more profitable and gives farmers an incentive to expand production onto additional land," suggesting that whether the predictions of the FT model are true or not depends to a great extent on specific sociopolitical and ecological circumstances. Examining 17 case studies from Latin America, Africa, and Asia, these authors conclude that the issue of intensification of agriculture and its relationship to deforestation is complex and, effectively, that agricultural policy could be modified in such a way as to promote forest-preservative policies rather than policies that, however unintentionally, actually promote more deforestation with "improved" agricultural technologies. Below, in the context of our matrix quality model, we discuss the qualitative nature of the sorts of agricultural development models that might be expected to restrain deforestation.

To be sure, a few studies show support for the forest transition model whereas others describe more complex situations, but the great majority of the studies show no effect or increased deforestation with either agricultural intensification or rural population decline. Other studies reflect similar complexity:

i.   In the Sarapiqui region of Costa Rica, in spite of all of the conditions appropriate for the FT model (agricultural intensification, a national shift to an industrial and service economy that attracts people from rural to urban areas) in addition to changes in attitude of landowners in favor of forests (in part due to an increase in ecotourism), forest recovery has been prevented and forest fragmentation has continued due to the concentration of land into absentee-owned cattle ranches, producing what has been called "hollow frontiers".

ii. In El Salvador, through analysis of satellite images, it was found that local rural population density was uncorrelated with forest recovery, whereas remittances from family members living abroad correlated positively with forest recovery.

iii. In a review of the evidence surrounding the claim that population drives deforestation in Panama, Sloan concludes that where institutional, economic, or contextual factors are considered, population-deforestation correlations are found to be "spurious or even counter-intuitive."

iv. In Missiones, Argentina, Izquierdo et al. note that although the population growth rate is slowing and the rural population is declining, forest cover continues to decline. They further note that, especially when soil and other physical conditions are not limiting, rural-to-urban migration does little to prevent further agricultural penetration into natural habitat, as has been happening in the Atlantic forest of Brazil.

v. A recent review of 17 studies of rural population dynamics in Mexico found little
    evidence that either intensification (in the form of eliminating peasant agricul-
    ture) or rural outmigration has had the result expected from the forest transition
    paradigm. Of the 17 studies, 16 exhibited net deforestation even though the back-
    ground conditions correspond to the requisites for the FT model to be applicable.

In sum, social context makes a difference in the direction as well as the degree of impact
of agricultural intensification on deforestation, what Schmink calls the "socioeconomic
matrix of deforestation". These and other studies reject the simplifying assumptions of
the forest transition model and echo the call of Angelsen and Kaimowitz for careful examination of the social-political forces operative in land-use planning so as to develop
programs that indeed will function to reduce deforestation. It is clear that the optimistic projections of a simple forest transition model, taking from the experience of some
regions (e.g., Eastern United States, Europe) and applying wholesale to tropical regions
in today's political climate, could be misleading.

In a break with such simplifying assumptions, Hecht proposes the addition of a new
conceptual framework specifically tuned to the contemporary situation (and most
evidently applicable to Latin America). This new conceptual framework is called the
"new rurality," and categorizes rural landscapes into four broad and overlapping
categories: environmental, socioenvironmental, agroindustrial, and peasant. Such a
categorization would not have made a great deal of sense either before the Cold War
or during the heydays of neoliberalism after the Cold War, but, argues Hecht, it is a
framework that strongly aids our understanding of rural dynamics in the contemporary world as it has been unfolding since the end of the Cold War. Analysts concerned
with rural landscapes tend to fall into one of these categories, and their analysis is
consequently driven by the vision they bring to the table. Environmentalists seek to
preserve native habitats, socioenvironmentalists seek to incorporate indigenous and
local communities in their conservation plan, and agroindustrialists see tremendous
opportunity in the expansion of industrial agriculture, which sometimes includes,
sometimes excludes the peasant element. Those who see the rural areas still populated with peasants (small family farms) see them acting in a variety of complex ways,
sometimes with strong economic and sociocultural links to cities. These complicated
actions and linkages ultimately will determine the fate of rural landscapes, according
to this point of view.

The Matrix Quality Model

Aligning ourselves effectively in Hecht's description of those who see rural areas still
populated with peasants and small-size family farms, and focusing on the past few decades of development in the science of ecology, we argue that data and theory suggest that conservation should be viewed from a larger landscape perspective and that,
with that perspective, moving agriculture toward a sustainability priority rather than a
productivist priority has more potential to affect biodiversity conservation positively.

Furthermore, there is at least circumstantial evidence that such a model would help, indirectly, to solve several aspects of the world food crisis.

Ecological Component: A Mean-field Approach

Reflecting older arguments in ecology, the standard preservationist attitude is effectively a "local carrying capacity" attitude, focusing on the size of a natural area, noting, correctly, that a minimum area is required for the long-term persistence of target species but failing to acknowledge up front that the larger landscape is sometimes more important for species survival than the size of a particular patch of natural habitat. This preservationist attitude has been criticized mostly from a social, moral, and ethical point of view. More recently, the criticism has been enriched with ecological theory that supports what might be called an "interfragment migration" approach, deriving mainly from recent ecological research on metapopulations. This new approach emphasizes the matrix within which fragments are located, and frames the argument as the "quality" of that matrix. This framing can be formalized through the use of metapopulation theory. To this end, an extension of the Levins model has been employed, namely, letting p be the proportion of potential habitats occupied by the species in question, m be the migration rate, and e be the extinction rate,

$$\frac{dp}{dt} = m(h-p)p - ep,$$

where h is the amount of appropriate habitat still available (h = 1 is an unperturbed habitat). Thus, the equilibrium situation will be p* = h − e/m, and the critical habitat loss that results in regional extinction would then be h = e/m.

This approach carries with it the critical assumption that as habitats are lost, the migration coefficient will remain constant. This assumption is not likely to be satisfied in many cases in nature. Consider, for example, a set of n very small habitat patches arranged in a one-dimensional space of length N. The average distance between patches is N/n. If the fraction of suitable habitat remaining is changed from 1 to h (where 0 < h < 1), the number of habitats will be hn, and the distance between patches will be approximately N/hn. Given an organism that is capable of migrating at some fixed rate, the effective patch-to-patch rate will be proportional to N/hn, which is to say the migration rate will be a function of h, the fraction of remaining suitable habitats. Thus, in $\frac{dp}{dt} = m(h-p)p - ep$, the migration coefficient should be replaced by a function of h.

As a first approximation, take the function to be a simple proportion (that is, the migration coefficient multiplied by the fraction of suitable habitat remaining = m_1h), which gives:

$$\frac{dp}{dt} = m_1 h(h-p)p - ep$$

with an equilibrium value of $p^* = h - e/m_1 h$, whence we can calculate that the metapopulation will persist (i.e., p^* will be greater than zero) as long as:

$$h > \sqrt{\frac{e}{m_1}}.$$

And because $e/m_1 < 1$ for persistence even without habitat destruction, we note that

$$\sqrt{\frac{e}{m_1}} > \frac{e}{m_1},$$

which means that the original notion that h must be greater than the extinction-to-migration ratio for persistence is optimistic. Because of the common, if not inevitable, reduction in overall migration rate with the reduction in fragment numbers, the critical habitat loss is scaled to the square root of that ratio, not the ratio itself.

From the point of view of our matrix quality model, an additional point about h is essential. In the real world it is only rarely the case that habitats are "completely" destroyed. Furthermore, a great deal of conservation biology now concerns itself with the quality of the matrix, partially because of the significant amounts of biodiversity that may be contained therein but especially because interfragment migration is necessary for metapopulation survival. In previous work, the limited nature of the classic metapopulation approach has been noted, especially with respect to its assumption that the matrix in which subpopulations are situated is homogeneous, showing one way in which that assumption could be relaxed—that is, by allowing the quality of the matrix to enter the basic equation as a linear input to the migration rate. The framework presented here expands on that relaxation by focusing on h, a focus explicitly relevant to anthropogenic landscapes but retaining the heuristic convenience of the mean-field approach. If h is the amount of original habitat left, suppose the rest is divided between q_1 and q_2, good-quality matrix and poor-quality matrix, respectively. Suppose the good-quality matrix (q_1) in fact does permit the same migration coefficient as when $h = 1$ but there is a significant reduction in the poor-quality matrix (q_2). Thus, assuming $q_2 = 0$, we have:

$$\frac{dp}{dt} = m_1 h (h + q_1)(h - p)p - ep$$

with an equilibrium value of $p^* = h - e/m_1(h + q_1)$, whence we calculate that the metapopulation will persist as long as:

$$h > \sqrt{\frac{q_1^2}{4} + \frac{e}{m_1}} - \frac{q_1}{2}.$$

And, comparing this value with the original criterion on h, we find persistence always enhanced by matrix quality, not surprisingly. Relating the critical habitat with zero matrix quality to the critical habitat with q_1 matrix quality, we can formulate the benefit of improving matrix quality as the ratio of those two critical habitats, which is:

$$\frac{1}{\sqrt{1 + \dfrac{m_1 q_1^2}{4e}}}$$

The somewhat surprising result that an improvement in matrix quality can outweigh the negative effects of habitat loss at values of $h > 1 - q_1$, a fact that could have important practical consequences and clearly relates to the question of what is being done in the matrix habitat. It is worth noting also that, as in the standard metapopulation model, when *p* is very small it is especially sensitive to changes in migration and extinction rates.

This approach, using the simple mean-field metapopulation model, only relates to the question of persistence or extinction of a particular species, and is, effectively, an extension of previous approaches. Scaling up to the community level is in the realm of metacommunity theory. If a metacommunity is thought of as only a collection of metapopulations (not the only possible definition), then our argument extends in an elementary fashion. Furthermore, we acknowledge the obvious fact that the direct biodiversity conservation value of agriculture varies greatly, with some forms of agriculture well-known to contain within their associated biodiversity almost as many species of some taxa as the natural habitat from which they were carved. Finally, we note that matrix quality will vary for different species, and in particular with the type of natural habitat that agriculture replaces.

Extinctions in Fragments and Migrations through the Matrix

Much of spatial ecological theory depends on extinction as one of the major processes driving patterns, including patterns of biodiversity. Although the fact of local extinctions is well-established it does not occur randomly, and certainly deserves more study. Nevertheless, there is little doubt that amid many complications, populations living in isolated fragments of natural vegetation can expect to experience extinctions, if enough time passes. If conservation is to be a long-term goal, this elementary and undeniable fact must be incorporated into planning.

A further complication may result from spatial self-organization. Consider, for example, plant communities in which the constituent species tend to expand in space through seed dispersal but are attacked by natural enemies in a density-dependent fashion according to the Janzen/Connell effect. It can be shown that such an arrangement will result in the clumping of organisms even in a uniform environment. Because of the dynamic interplay of seed dispersal and density-dependent control, any given clump is expected to go locally extinct over the long run. In such a situation, fragmenting the continuous habitat does not change much about the local extinct rates, which are a consequence of density-dependent operation of natural enemy dynamics. However, normal migration (i.e., seed dispersal) will be reduced.

Unfortunately, long-term studies that uncover such patterns of extinctions in continuous habitat are not common in the literature. Rooney et al. demonstrated dramatic changes in species composition in plots embedded in natural forest communities in the northern Great Lakes region of the United States. Environmental drivers in this case included forces such as deer hunting and invasive species, but one of their key results is that, even in this unfragmented forest, species loss at a local level was dramatic. In a 20-year study of the amphibians occupying small ponds in a forested matrix, $\approx$30 local extinction events were observed. In this case, the researchers were able to demonstrate that "reinvasions," which is to say, migration events, completely balanced these local extinctions. In summary, both ecological theory and empirical studies strongly suggest a three-part conclusion. First, local extinctions are normal and occur even in areas of continuous natural habitats. Second, migrations throughout the matrix can balance those extinctions and maintain a metapopulation structure that will prevent regional extinction. Third, the quality of the matrix matters; high-quality matrices are those that promote migration, thus maintaining metapopulation structures that obviate regional extinction.

Convergence of Food Production with Nature Conservation

The matrix quality model challenges the assumption that agriculture is the enemy of conservation. It is the kind of agriculture, not the simple fact of its existence that matters. Whether looked at from the point of view of the simple mean-field model

$$(\frac{dp}{dt} = m(h-p)p - ep, h > \sqrt{\frac{e}{m_1}}. \ h > \sqrt{\frac{q_1^2}{4} + \frac{e}{m_1} - \frac{q_1}{2}}.)$$ or from the more qualitative em-

pirical fact that some habitats promote more migration than others, the agricultural matrix is perhaps the most important habitat on which conservation efforts must focus. But this brings us face to face with one of the multiple functions of agriculture: to produce food, fiber, drugs, and energy for human use.

Regarding the productivity of agriculture, we face what seems at first to be a dilemma. The sort of high-energy-demanding, chemically intensive agriculture associated with modernity generates a very low quality matrix, whereas alternative agriculture (organic, agroecological, natural-systems agriculture, etc.) would seem to be precisely the forms that would produce a high-quality matrix. Yet it is just such agricultural types that are normally assumed to be less productive. A simple accounting from this assumption is precisely what generates the land-sparing model, the forest transition model, and the optimistic assessments that rural-urban migration, as it decreases the number of "peasant" producers (automatically presumed to be inefficient), will result in equivalent, or even higher, production on less land, generating more forest recovery. However, what evidence supports this fundamental assumption.

Anecdotes can easily support the assumption, especially when highly subsidized farmers from the United States and other industrialized regions are compared with small-scale farmers of the Global South, and the measure of productivity is yield of the main commercial crop or net profit. However, if the measure of productivity is simply total output per area, relevant data do not seem to support the basic assumption. For example, analyzing data relating farm size to productivity (output per unit area), Cornia found that in all cases the trend was decreasing productivity as farm size increased. Indeed, the "productivity-size inverse relationship" is a well-known fact among agricultural economists, and was first pointed out by Nobel laureate Amartya Sen in the 1960s. It seems that small owner-operated farms tend to be more efficient in that the farmer knows the land and its ecology well, and plants crops with that knowledge, usually using a multicropping strategy to take advantage of local peculiarities such as, for example, the Kayapó's management of their Amazonian landscape where the patches of the matrix are an entangled mosaic that takes advantage of microclimatic and soil differences to produce and promote hundreds of species of plants and animals. Many other examples could be cited. Contrarily, large, highly capitalized farms seek economies of scale in which those local ecological peculiarities are purposefully ignored. Ironically, the recent enthusiasm for so-called precision farming acknowledges precisely this underlying ecological structure, but proposes to resolve it with a high-tech strategy of sensors and delivery systems. As one of our students reviewing the literature on precision farming quipped, "Small-scale farmers already do precision farming." Thus, both the logic and the data suggest that small-scale agriculture can be more productive, on a per unit-area basis, than large-scale agriculture.

The assumption that large-scale intensive monocultures are more productive than agroecological and organic systems is likewise debatable. In a recent review of almost 300 studies comparing yields of organic/agroecological and conventional agriculture throughout the world, it was found that, on average, organic and agroecological systems produce as much, if not more, than conventional systems, corroborating many other studies. Furthermore, it has now been well-established that energy efficiency in traditional and many organic systems is higher than in high-industrial/conventional agriculture.

In summary, contrary to the conventional wisdom that industrial agriculture is needed to produce enough food to feed the world, the empirical evidence suggests that peasant and small-scale family farm operations adopting agroecological methods can be as (or more) productive than industrial agriculture. Given that most of the world's poor live in rural areas or are urban poor recently displaced from rural areas, an agricultural matrix composed of small-scale sustainable farms can thus create a win-win situation that addresses both the current food crisis and the biodiversity crisis.

However, there exists a very complicated irony that is rarely addressed. The search for more productivity, part and parcel of the research agenda of most agricultural researchers, is not necessarily a rational project. In many cases (and here coffee and maize would be excellent recent examples), the major agricultural problem is "overproduction" and consequent low prices. The recent (and temporary) increase in food prices notwithstanding, it is often the case that farmers receive inadequate compensation for their efforts largely because markets become saturated. If unregulated markets must be the rule, an assumption that itself might be questioned, overproduction and low prices will continue to plague farmers, not continuously, but on a boom-and-bust cycle. Indeed, the IAASTD, an intergovernmental assessment process that involved 3 years of research and 400 experts from all over the world, concluded that conventional/industrial agriculture is not a rational option for alleviating poverty and ending hunger and malnutrition nor for sustainable development, further noting that more equality is needed for alleviating hunger and malnutrition. This equality is more likely to be achieved through a land reform that redistributes land that is in the hands of big agrobusiness and planted in commercial monocultures and puts it in the hands of small- and medium-size family farmers who are more likely to construct a landscape mosaic that promotes biodiversity and produces more food.

Agroecological Farming as Key to a Sustainable Future

More than 220 million people in Sub-Saharan Africa do not have enough to eat, and nearly one in four are undernourished. With population growth outpacing food production, and sixty percent or more of the region's population depending on agriculture for food and income, those numbers are growing.

Rising temperatures and drought, among other extreme weather events, already threaten food production for some of the world's most climate-vulnerable countries. Improving agricultural output of smallholders through agroecological farming could be massively beneficial in delivering the United Nations Sustainable Development Goal to end world hunger, achieve food security and improve nutrition, and promote sustainable agriculture.

What does Agroecology offer Farmers?

Agroecology addresses the root causes of hunger, poverty and inequality by helping to transform food systems and build resilient livelihoods through a holistic approach that balances the three dimensions of sustainability – social, economic and environmental. Agroecological agriculture (of which organic is one system) supports small farms that are diverse, integrated and use low levels of input to ensure the long-term balance between food production and the sustainability of natural resources.

A number of international organisations and African NGOs argue that agroecology should be the future of agriculture on the continent. But broader adoption requires training and support for farmers to embrace the approach, instead of relying on the short-term convenience of expensive chemical inputs.

Wider Benefits of Agroecological Farming in Sub-Saharan Africa

A 2008 UN study on the productivity performance of organic and "near organic" agriculture in Africa found that average crop yields increased by 116 percent (128 percent in East Africa specifically), with a corresponding increase in household food security.

Send a Cow's projects highlight that agroecology's benefits extend far beyond food production. A training program in Uganda helped increase the education level of participant's children by 145 percent. In Kenya, female decision-making in the home increased by 147 percent while simultaneously raising earning power of women.

Agroecological techniques can improve the resilience of farming systems by increasing diversification through poly-cropping, agroforestry, integrated crop and livestock systems, and the use of local varieties. This resilience can reduce the risks of pests and diseases and the costs of seeds. The management of soil fertility through rotations, cover crops and manuring can increase soil water retention or drainage, offer a better response to droughts and floods, reduce the need for irrigation, and help avoid land degradation.

Moreover soil quality is improved with higher levels of organic matter, which helps mitigate climate change by sequestering carbon in the soil.

Agroecology and Sustainable Food Production System

While agriculture was very successful on meeting global food demands in the 20th century, now the threats and impact of the practices and policies followed raises the need for a paradigm shift towards a truly sustainable food production system. A sustainable food production system is a collaborative network that integrates several components in order to enhance a community's environmental, economic and social well-being. It is built on principles that further the ecological, social and economic values of a community and a region.

Agroecology uses disciplines from modern agricultural science but its approach is also influenced by the indigenous knowledge systems about soils, plants, etc. that have nurtured traditional farming systems for millennia. Agroecology does not promote technical recipes but rather principles, it is not an agriculture of inputs but rather of processes.

In order for the technologies derived from the application of principles to be relevant to the needs and circumstances of small farmers, the technological generation process must result from a participatory research process in which farmers along with researchers provide input into the research questions and the design, running and evaluation of field experiments. Besides science and practice, agroecology also refers to a wide variety of social movements aimed at environmental protection, the development of sustainable farming systems and food sovereignty.

The concept of "movement" is used in order to stress the vision of social and economic positive impact that Agroecology potentially has for sustainable rural development The Agroecology sector emerged as a different model to address the problems of world agriculture caused by industrial food production model and it is based on the principles of Sustainability, Integrity, Equality, Performance and Stability.

The primary concepts of agroecology and corresponding management practices resonate with arguments for food security, food sovereignty and sustainable rural development. The agroecological concepts and principles embrace also a wide range of practices and have broad scope for implementation. This means that they have considerable resonance with other concepts, principles and practices in the field of sustainable agriculture that also offer alternative structures to the mainstream paradigm of industrial agriculture. Such key approaches are Permaculture, biodynamic agriculture, organic farming, conservation agriculture, urban farming, and multifunctionality in agriculture.

The context of agroecology expanded over the years from the field, through the farm, to the landscape; and in a wider understanding to the food system level. Agroecology promotes ecologically and culturally sound food systems and food sovereignty, protecting people's ability and right to define their own models of food production, distribution and consumption. Such an approach in rural development can contribute to empowering disadvantaged communities through diversifying activities of farmers and including new groups in different levels of the food system- production, processing and trade. This helps to strengthen employment; local food security and food and prevent population decrease.

References

- Agroecology: arc2020.eu, Retrieved 4 March, 2019

- What-is-regenerative-agriculture: regenerationinternational.org, Retrieved 11 July, 2019

- Carbon-farming: carboncycle.org, Retrieved 21 February, 2019

- Permaculture: e-csr.net, Retrieved 1 March, 2019

- Agroforestry-and-its-benefits: reset.org, Retrieved 25 June, 2019

- Agroecology-Scaling-up-agroecology: fao.org, Retrieved 6 January, 2019

- Agroeco-principles: agroeco.org, Retrieved 16 August, 2019

- Green-Revolution: people.ku.edu, Retrieved 7 April, 2019

- Agroecological-farming-key-sustainable-future: agrilinks.org, Retrieved 18 January, 2019

- Introductions-concept-agroecology-sustainable-food-production-system: rusdelaproject.eu, Retrieved 20 May, 2019

Chapter 2

Agro-ecological Land Resources Assessment

Some of the aspects which are dealt with under agro-ecological land resources assessment are agro-ecological zones assessment, agro-climatic suitability classification and agro-edaphic suitability classification. The chapter closely examines these key techniques of agro-ecological land resources assessment to provide an extensive understanding of the subject.

Agro-ecological Zones Assessment

The Agro-ecological Zones approach is a GIS-based modeling framework that combines land evaluation methods with socioeconomic and multiple-criteria analysis to evaluate spatial and dynamic aspects of agriculture.

The Food and Agriculture Organization of the United Nations (FAO), in collaboration with the International Institute for Applied Systems Analysis (IIASA), has developed the Agro-ecological Zones (AEZ) methodology and a worldwide spatial land resources database. Together this enables an evaluation of biophysical limitations and production potential of major food and fiber crops under various levels of inputs and management conditions.

When evaluating the performance of alternative types of land use, a single criterion function often does not adequately reflect the decision-maker's preferences, which are of a multiple-objective nature in many practical problems dealing with resources planning. Therefore, interactive multiple-criteria model analysis has been introduced and applied to the analysis of AEZ models. It is at this level of analysis that socioeconomic considerations can effectively be taken into account, thus providing a spatial and integrated ecological–economic planning approach to sustainable agricultural development.

Future land uses and agricultural production are not known with certainty. For example, what will be the availability and adoption of agricultural technology for various crops in the future? What new genetic crop varieties will be available? How will climate change affect crop areas and productivity? A scenario approach based on a range of assumptions related to such changes in the future enables assessments and a distribution of outcomes that facilitate policy considerations and decision making in the face of future uncertainty.

The AEZ approach, estimated by grid cell and aggregated to national, regional, and global coverage, provides the basis for several applications. These include the following:

- Identification of areas with specific climate, soil, and terrain constraints to crop production.

- Estimation of the extent of rain-fed and irrigated cultivable land and potential for expansion.

- Quantification of crop productivity under the assumptions of three levels of farming technology and management.

- Evaluation of land in forest ecosystems with cultivation potential for food crops.

- Regional impact and geographical shifts of agricultural land and productivity potentials and implications for food security resulting from climate change and variability.

Methodology

The AEZ methodology follows an environmental approach: it provides a standardized framework for the characterization of climate, soil, and terrain conditions relevant to agricultural production. Crop modeling and environmental matching procedures are used to identify crop-specific environmental limitations under assumed levels of inputs and management conditions. The elements involved in the AEZ framework are described in figure.

FAO's Digital Soil Map of the World has been made the reference for constructing a land surface database consisting of more than 2.2 million grid cells at 5- minute latitude/longitude within a raster of 2160 rows and 4320 columns. On the input side, the key components of the database applied in the AEZ methodology include the following:

- The FAO Digital Soil Map of the World and linked soil association and attribute databases.

- The Global 30 arc-second Digital Elevation Model was used for elevation and the derived slope distribution database.

- The global climate data set of the Climate Research Unit of the University of East Anglia (CRU) consisting of average data (for the period from 1961 to 1990) and data for individual years from 1901 to 1996.

- A layer providing distributions in terms of 11 aggregate land-cover classes derived from a global 1-kilometer land-cover data set.

The AEZ global land resources database also incorporates spatial delineation and accounting of forest and protected areas. A global population data set for the year 1995 provides estimates of population distribution and densities at a spatially explicit subnational level for each country.

On the output side, numerous new data sets have been compiled at the grid-cell level and tabulated at the national and regional levels. Outputs include:

- Agro-climatic characterizations of temperature and moisture profiles, and

- Time series of attainable crop yields for all major food and fiber crops.

The AEZ methodology considers the contribution of multiple cropping to land productivity on the basis of the evaluation of thermal and moisture profiles in a grid cell for determination of agronomically meaningful sequential crop combinations.

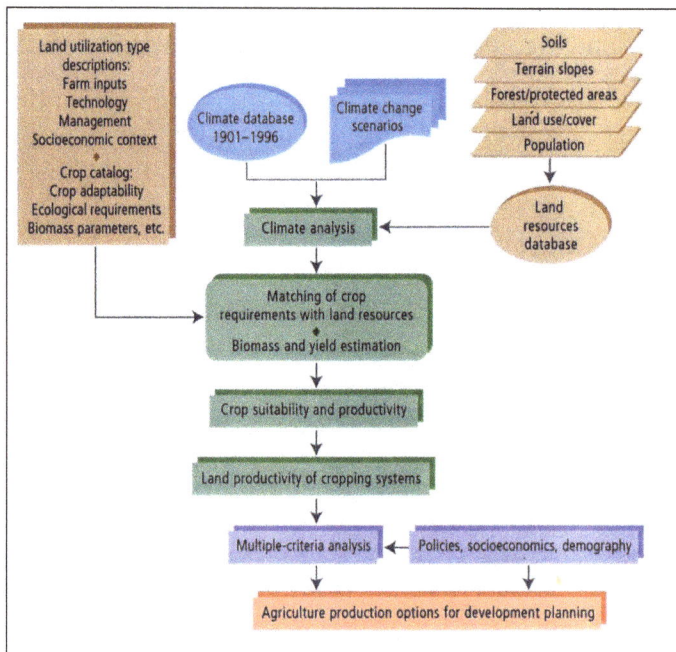

Agro-Ecological Zones (AEZ) methodology

The AEZ framework incorporates the following basic elements:

- Selected agricultural production systems with defined input and management relationships, and crop-specific environmental requirements and adaptability characteristics. These are termed "land utilization types" (LUTs). The AEZ study distinguishes some 154 crops, fodder, and pasture LUTs, each at three levels of inputs and management (high, intermediate, low).

- Geo-referenced climate, soil, and terrain data, which are combined into a land resources database. The computerized global AEZ database contains some 2.2 million grid cells.

- Accounting for spatial land use and land cover, including forests, protected areas, population distribution and density, and land required for habitation and infrastructure.

- Procedures for calculating the potential agronomically attainable yield and for matching crop and LUT environmental requirements with the respective environmental characteristics contained in the land resources database, by land unit and grid cell.

- Assessment of crop suitability and land productivity of cropping systems.

- Applications for estimating the land's population supporting capacity, multiplecriteria optimization incorporating socioeconomic and demographic factors of land resource use for sustainable agricultural development.

The AEZ assessments were carried out for a range of climatic conditions, including a reference climate with data on individual historical years, as well as scenarios of a future climate based on various global climate models. Farming technology was considered at three levels: a high level of inputs with advanced management, an intermediate level with improved management, and a low level of inputs with traditional management. Hence, the results quantify the impacts on land productivity of both historical climate variability and potential future climate change.

Intensity level	Characteristics
High Level of Inputs/Advanced Management	Production is based on improved high-yielding varieties and is mechanized with low labor intensity. It uses optimum applications of nutrients; chemical pest, disease, and weed control; and full conservation measures. The farming system is mainly market oriented.
Intermediate level of Inputs/Improved Management	Production is based on improved varieties and on manual labor and/or animal traction and some mechanization. It uses some fertilizer application and chemical pest, disease, and weed control, and employs adequate fallow periods and some conservation measures. The farming system is partly market oriented.
Low Level of Inputs/Traditional Management	Production is based on the use of traditional cultivars (if improved cultivars are used, they are treated in the same way as local cultivars) and labor-intensive techniques, with no application of nutrients. It uses no chemicals for pest and disease control and employs adequate fallow periods and minimum conservation measures. The farming system is largely subsistence based.

The AEZ results indicate that, at the global level, Earth's land, climate, and biological resources are ample to meet food and fiber needs of future generations, in particular, for a world population of 9.3 billion, as projected in the United Nations medium variant for the year 2050. Despite this positive aggregate global picture, however, there are reasons for profound concern in several regions and countries with limited land and water resources.

Socioeconomic development will inevitably infringe on the current and potential agricultural land resource base, as the need to expand industrial, infrastructure, and habitation land use increases. Furthermore, global environmental changes, particularly climate change, are likely to alter the conditions and distribution of land suitability and crop productivity in several countries and regions.

The presentation of results is organized as follows:

- Climate, soil, and terrain limitations to crop production.

- Land with cultivation potential.

- Potential for expansion of cultivated land.

- Cultivation potential in forest ecosystems.

- Yield and production potentials.

- Temperature and rainfall sensitivity.

It should be noted that the AEZ results have been aggregated to the national, regional, and global levels. Furthermore, the farming technology and input assumptions are based on present-day knowledge. Research and scientific developments in the future could alter the projection outcomes.

Climate, Soil and Terrain Limitations to Crop Production

Climate constraints are classified according to the length of periods with cold temperatures and moisture limitations. Temperature constraints are related to the length of the temperature growing period, i.e. the number of days with a mean daily temperature above 5 °C. For example, a temperature growing period shorter than 120 days is considered a severe constraint, while a period shorter than 180 days is considered to pose moderate constraints to crop production. Hyper-arid and arid moisture regimes are considered severe constraints, and dry semi-arid moisture regimes are considered moderate constraints.

Soil constraints are classified into moderate and severe limitations imposed by soil depth, fertility, and drainage; soil texture/structure/stoniness; and specific soil chemical conditions. Limitations imposed by terrain slope have been classified similarly. The extent of land with climate and soil/terrain constraints is shown in below figure.

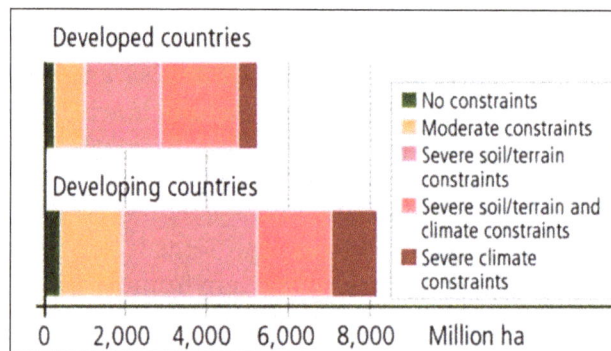

Extent of land with climate and soil/terrain constraints.

On the basis of currently available global soil, terrain, and climate data, the AEZ estimates indicate that 10.5 billion hectares (ha) of land—more than three-quarters of the global land surface, excluding Antarctica—suffer rather severe constraints for rainfed crop cultivation. Some 13% of the surface is too cold, 27% is too dry, 12% is too steep, and about 65% is constrained by unfavorable soil conditions, with multiple constraints coinciding in some locations. Figure shows the distribution of land constraints by region, and figure portrays the situation worldwide.

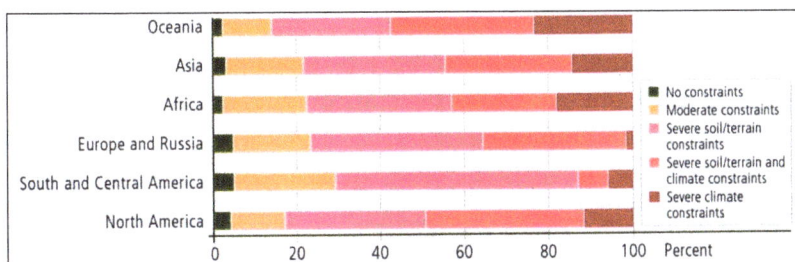

Distribution of climate and soil/terrain constraints by region.

At the global level, almost 40% of the soils suffer from severe fertility constraints and about 6% are affected by limitations resulting from salinity, sodicity, or gypsum constraints. The respective regional figures are 43% and 1% for North America; 46% and 5% for South and Central America; 56% and 4% for Europe and Russia; 30% and 3% for Africa; 28% and 11% for Asia; and 31% and 18% for Oceania.

Climate change is likely to have both positive and negative effects on extent and productivity of arable land resources. In some areas, prevailing constraints may be somewhat relieved by climate change, thus increasing the arable land resources. In other areas, however, currently cultivated land may become unsuitable for agricultural production.

The extent to which specific constraints like low fertility and toxicity can be overcome will also depend on the outcomes of agricultural and scientific research. For example, agricultural research in Mexico has resulted in the application of biotechnology to increase plant tolerance to aluminum, thus countering soil toxicity problems common in some tropical areas.

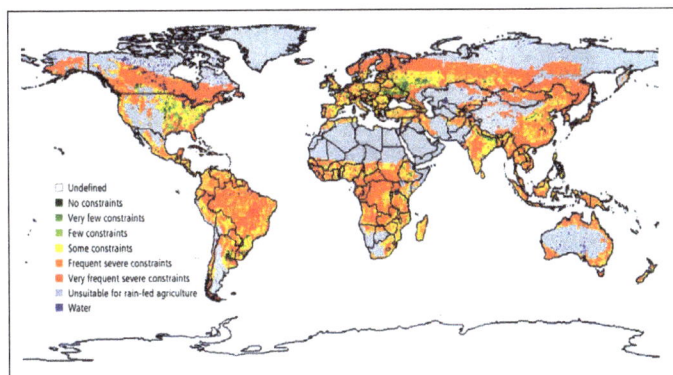

Climate and soil/terrain constraints combined at worldwide level.

Land with Cultivation Potential

There are various ways to estimate the extent of land with cultivation potential for rainfed crops. Any quantification depends on a variety of assumptions concerning the range of crop types considered; the definition of what level of output qualifies as acceptable; and the social acceptance of land-cover conversions (of forests in particular) and what land constraints may be alleviated with farming technology, management, and investment.

The AEZ assessment considers a total of 24 crop species, two pasture types, and two fodder crops. Altogether, 154 crop/land utilization types are considered, each at three defined levels of inputs and management.

The results of the estimation of extent of land with cultivation potential for major crops show that at the global level about 2.7 billion ha are suitable for cereal cultivation with a high level of inputs and management, and about three-quarters of this land is very suitable or suitable. The land area suitable for roots & tubers and pulses is some 30% to 40% smaller than that suitable for cereal.

Table: Extent of land with rain-fed cultivation potential for major crop groups (million ha).

Level of inputs	Suitability of land	Cereals	Roots & tubers	Pulses	Oil crops	Sugar crops	Cotton
High	VS + S	2,000	1,145	1,050	1,495	535	440
	VS+S+MS	2,725	1,865	1,720	2,410	1,150	755
Intermediate	VS+S	1,590	795	725	1,105	385	280
	VS+S+MS	2,500	1,515	1,495	2,035	900	580
Intermediate excluding forests, land for settlement and infrastructure	VS+S	1,255	595	645	890	270	270
	VS+S+MS	1,960	1,130	1,255	1,585	595	550

VS=very suitable; S=suitable; MS=moderately suitable

Table also shows that the cultivable land for each of these major groups of crops is reduced by 10% to 20% at the intermediate level of inputs compared with the high level of inputs. The exclusion of land that is in forest ecosystems and land used for settlement and infrastructure would further reduce the cultivable land area for each of these crop groups by a similar proportion.

It should be emphasized that these results indicate the potential area suitable for each of these individual crop groups. In reality, the demand mix for domestic consumption and trade will drive allocation of land to particular crops.

The total extent of potential rain-fed land is estimated for each grid cell. When considering all modelled Global AEZ crop types excluding silage maize, forage legumes, and grasses, mixing all three input levels, and assuming no restrictions for land-cover conversion, the results show that about one-quarter of the global land surface, excluding Antarctica, can be regarded as potentially suitable for crop cultivation.

The total extent of land suitable for at least one crop amounts to some 3.3 billion ha. Of this, about 23% are in land classified as forest ecosystems. If only the very suitable and suitable land area is considered, then the corresponding extent of land is 2.5 billion ha, with some 24% in forest ecosystems.

Table: Rain-fed cultivated land in 1994–1996 and rain-fed cultivation potential for major food and fiber crops, mixed inputs (million ha).

Region	Total land	Cultivated land 1994–1996		Land with cultivation potential				Settlements and infrastructure
				VS+S		VS+S+MS		
		Rain-fed	Irrigated	Total	In forest ecosystems	Total	In forest Ecosystems	
Oceania	850	50	3	76	12	116	17	1
Asia	3,113	376	180	406	36	516	47	83
Africa	2,990	185	12	767	114	939	132	26
Europe & Russia	2,259	289	25	328	61	511	97	21
South & Central America	2,049	141	18	697	281	858	346	16
North America	2,138	203	22	266	96	384	135	9
Developing countries	8,171	702	208	1,872	433	2,313	527	124
Developed countries	5,228	543	53	669	168	1,012	247	33
World	13,400	1,245	260	601	601	3,325	774	156

VS=very suitable; S=suitable; MS=moderately suitable.

In developed countries, about one-fifth of the total land has rain-fed cultivation potential. In developing countries, this proportion is slightly less than 30%. The estimate of cultivable rain-fed land potential is more than twice the area reported by the FAO as land actually in use for cultivation in 1994–1996. These AEZ estimates are high, since all land suitable for at least one crop is included in the assessment and crop choice is not constrained by current demand mix. Figure displays the land suitable for rain-fed crops across the world.

For each of the approximately 2.2 million grid cells of the database suitability results were calculated for each crop/LUT. The outcomes were mapped by means of a suitability index (SI). This index reflects the suitability make-up of a particular grid cell. In this index VS represents the portion of the grid cell with attainable yields that are 80% or more of the maximum potential yield. Similarly, S, MS, and mS represent portions of the grid cell with attainable yields 60%–80%, 40%–60%, and 20%–40% of the maximum potential yield, respectively. SI is calculated using the following equation: $SI = VS*0.9 + S*0.7 + MS*0.5 + mS*0.3$.

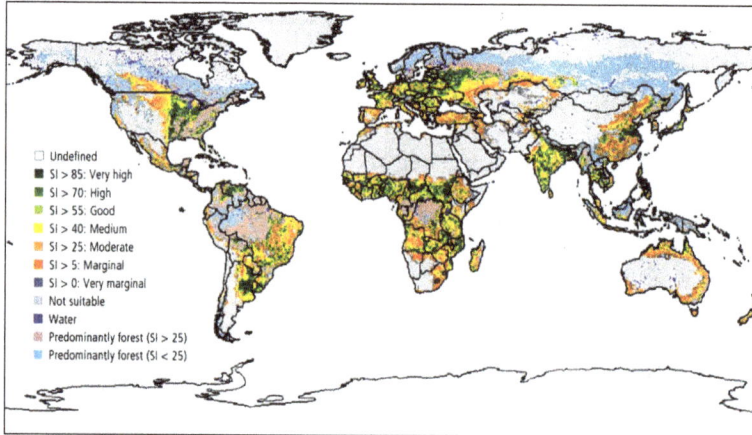

Suitability for rain-fed crops excluding forest ecosystems.

The results highlight that in Asia, Europe, and Russia, the rain-fed land that is currently cultivated amounts to about 90% of the potential very suitable and suitable land. Hence, there is little room for agricultural extensification. In the case of North America, some 75% of the very suitable and suitable land is currently under cultivation. By contrast, Africa and Latin America are estimated to have some 1.1 billion ha of land in excess of currently cultivated land; of this, about 36% is in forest ecosystems. In these two regions there is clearly scope for further expansion of agricultural land, even assuming that current forests are maintained.

Potential for Expansion of Cultivated Land

Despite the fact that currently reported cultivated land in official statistics is likely to underestimate actual use by some 10 to 20% in several developing countries, the results indicate that there is still a significant potential for expansion of cultivated land in Africa and South and Central America. More than 70% of additional cultivable (very suitable, suitable, and moderately suitable) land is located in these two regions, and about half of this land is concentrated in just seven countries—Angola, Democratic Republic of Congo, Sudan, Argentina, Bolivia, Brazil, and Colombia. In other regions, this potential is either very limited, as in Asia, or is unlikely to be used for agriculture in the future, as is the case in Europe and Russia, North America, and Oceania.

Agronomic suitability is by no means the only determinant of future land development. The potential expansion of cultivable land will be limited by the constraints of ecological fragility, degradation, toxicity, and incidences of diseases, as well as by a lack of infrastructure and limited financial resources. These issues will need to be considered explicitly in agricultural intensification at the national level.

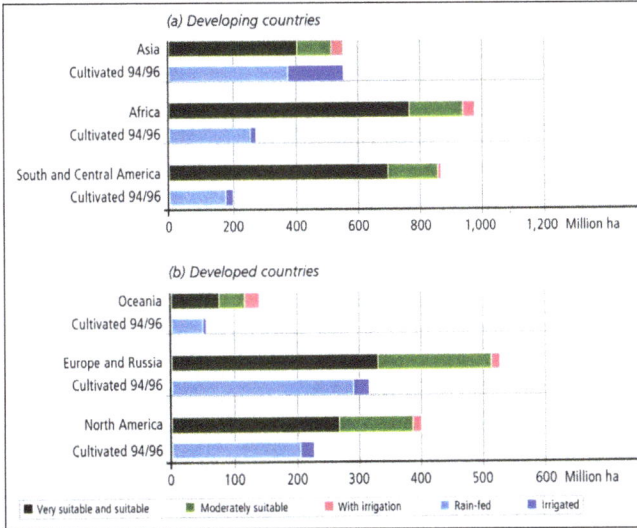

Comparison of land with crop production potential and land used for cultivation in 1994–1996.

Cultivation Potential in Forest Ecosystems

The FAO Global Forest Resource Assessment 2000 has mapped land area currently in closed forest ecosystems to be 2.9 billion ha, occupying some 21% of the world's land area. In addition, there are some 1.5 billion ha in open and fragmented forest ecosystems.

Russia and Brazil, which account for 19% of the total land area and 5.2% of the world population, have 36% of the global forest areas. The USA and Canada have some 470 million ha of forest, equivalent to 12% of the global forest areas. China, Australia, the Democratic Republic of Congo, and Indonesia account for a further 15% of forestland. These eight countries account for more than 60% of the total forest area in the world.

During the past decade, 127 million ha of the world's forest areas were cleared, while some 36 million ha of forest were replanted. China reported replanting about 50% of this gain. Europe, Russia, and the USA accounted for 24% of it. But Africa lost about 53 million ha of forest during this period.

The loss of forestland in most countries is primarily due to the expansion of crop cultivation. In some areas this is the result of population pressure for more arable land. In other places, like Brazil, commercial agriculture and livestock production are encroaching on forestland. Commercial logging of timber has also contributed to loss of forestland.

In the AEZ assessment, the cereal cultivation potential in forest areas reveals that some 470 million ha of land with cultivation potential for wheat, rice, or maize coincide with land classified as predominantly forest ecosystems, and this accounts for some 19% of land suitable for at least one of the three cereals.

Table: Land with rain-fed cultivation potential for wheat, rice, or maize in forest ecosystems (million ha).

Region	Total land	Total land in forest ecosystems	Land with rain-fed cultivation potential			
			VS+S		VS+S+MS	
			Total	In forest ecosystems	Total	In forest ecosystems
Oceania	850	72	44	7	298	11
Asia	3,113	388	263	14	384	25
Africa	2,990	246	404	25	592	43
Europe & Russia	2,259	761	282	41	463	76
South & Central America	2,049	751	283	128	474	200
North America	2,138	562	235	82	342	115
Developing countries	8,171	1,405	1,076	166	1,574	266
Developed countries	5,228	1,381	565	132	884	204
World	13,400	2,786	1,612	298	2,429	470

VS=very suitable; S=suitable; MS=moderately suitable.

Rather wide variations occur between regions. In Russia, for example, less than 9% of the land predominantly in forest ecosystems was assessed to have cultivation potential for cereal crops. Yet, this equates to about a quarter of Russia's land with rain-fed cultivation potential. In South America, these figures are 27% and 35%, respectively, and in North America they are 20% and 39%, respectively.

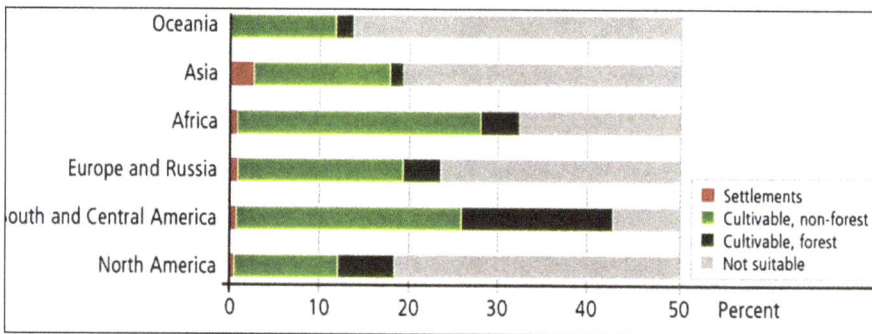

Share of total land suitable for crops, by forest and non-forest area.

Considering only the most suitable land in forest ecosystems, about 298 million ha are classified as very suitable or suitable for cultivation of wheat, rice, or maize with mixed levels of inputs. About 44% of this land is located in South and Central America, and altogether some 56% is located in the developing countries.

The spatially quantified information on the productive value of forest areas in terms of (1) potential values of crop and wood production, (2) the conservation role of forests in watershed management and flood control, and (3) their importance as carbon sinks and as habitats of rich biodiversity, is relevant in national considerations concerning the use and conservation of forest areas. These issues need to be central in international negotiations on preservation of regional forest ecosystems in the world.

Yield and Production Potentials

The maximum attainable yields in the AEZ assessment represent average values from simulated year-by-year agro-climatic yields during the period from 1960 to 1996. These yields were calculated for rain-fed and irrigated cultivation in the tropics, subtropics, and temperate/boreal zones.

Table: Maximum attainable yields under rain-fed and irrigated conditions (tons/ha)

	Input level	Region	Crop		
			Wheat	Rice	Maize
Rain-fed	Low	Tropics	2.7	5.0	5.1
		Subtropics	4.3	4.7	5.8
		Temperate	4.9	4.9	5.3
	Intermediate	Tropics	5.7	7.7	8.5
		Subtropics	8.4	7.3	8.9
		Temperate	8.7	6.9	8.7
	High	Tropics	8.5	9.9	12.5
		Subtropics	11.8	9.2	12.3
		Temperate	12.1	8.6	12.1
Irrigated	Intermediate	Tropics	7.4	9.5	10.5
		Subtropics	10.2	9.9	12.2
		Temperate	9.7	8.7	11.3
	High	Tropics	11.1	12.2	15.6
		Subtropics	14.2	12.7	17.1
		Temperate	13.5	10.9	15.7

Long-term yields, also estimated using AEZ procedures, assume that proper cycles of cropping and fallow are respected. At high levels of inputs with balanced fertilizer applications and proper pest and disease management, only limited fallow periods will be required. At low levels of inputs—assuming virtually no application of chemical fertilizer and only limited organic fertilizer, and very limited or no application of biocides—considerable fallow periods are needed in the crop rotations to restore soil nutrient status and to break pest and disease cycles.

The yields attained in the long term, when accounting for fallow period requirements, are well below the estimated short-term maximum attainable yields. Required fallow periods vary with soil and climate conditions. On average, long-term yields are 10%,

20%, and 55% lower than maximum attainable yields at high, intermediate, and low levels of inputs, respectively. It should also be noted that there is a more than threefold increase in short-term attainable yield and a sevenfold increase in long-term attainable yield in all regions as farming technology and management increases from low to intermediate levels of inputs. At the high level of inputs, the yield level increases further—by 60% to 80%. These estimates for low, intermediate, and high levels of inputs reflect present knowledge and technology.

Intensification of agriculture will be the main means to increase production. In many developing countries, provided adequate inputs and improved management are applied, there is considerable scope for increased yields. For example, in the developing countries the 1995–1997 actual yields per ha for wheat averaged about 1.8 tons for rainfed conditions and 3.1 tons with irrigation, with an overall average of 2.4 tons per ha, compared with 3.1 tons per ha for major developed country exporters.

Currently there is a wide variation in the level of inputs (e.g. fertilizer application) across regions. In sub-Saharan Africa, an average of 8 kg of nutrients are applied per ha, whereas in other developing countries the rate is about 80 kg of nutrients per hectare, and in the developed countries over 200 kg of nutrients are applied per hectare. These figures are overall averages and include both rain-fed and irrigated crop production. There is considerable scope for improved management and use of inputs—particularly of nutrients—in many developing countries, especially in Africa and South America, where the levels of application are low.

Table: Maximum short-term attainable and long-term sustainable yields for rain-fed wheat, rice, or grain maize averaged over all VS+S+MS land, by region and level of inputs (tons/ha).

Region	Low inputs		Intermediate inputs		High inputs	
	Short-term attainable	Long-term sustainable	Short-term attainable	Long-term sustainable	Short-term attainable	Long-term sustainable
Oceania	0.7	0.4	3.2	2.6	5.3	4.8
Asia	1.1	0.5	3.7	3.0	6.2	5.6
Africa	1.1	0.4	3.8	3.1	6.7	6.0
Europe & Russia	0.9	0.4	3.6	2.9	5.5	5.0
South & Central America	1.2	0.6	3.6	3.0	5.6	5.1
North America	0.8	0.4	3.6	2.8	5.8	5.2
Developing countries	1.1	0.5	3.7	3.0	6.2	5.6
Developed countries	0.9	0.4	3.5	2.8	5.6	5.1
World	1.0	0.5	3.7	3.0	6.0	5.4

VS=very suitable; S=suitable; MS=moderately suitable.

The environmental implications of increasing fertilizer and chemical inputs in the future will have to be taken into account, along with the lessons and experiences from the green revolution. Also, any strategy for increasing food production through intensification must consider the socioeconomic issues of small and resource-poor farmers—in particular, their access to and ability to purchase inputs.

In areas where growing periods are sufficiently long, the AEZ methodology takes into account viable sequential cropping. To perform this estimation, a multiple-cropping zone classification is used to determine feasible crop combinations. The algorithms used for constructing cropping patterns have been designed to ensure that typical crop sequences in cultivation cycles are used. For instance, in the typical double-cropping areas around Shanghai in China, rice or maize was selected as the most productive summer crop, and wheat or barley was chosen as the winter crop. Figure shows the occurrence of multiple-cropping zones worldwide.

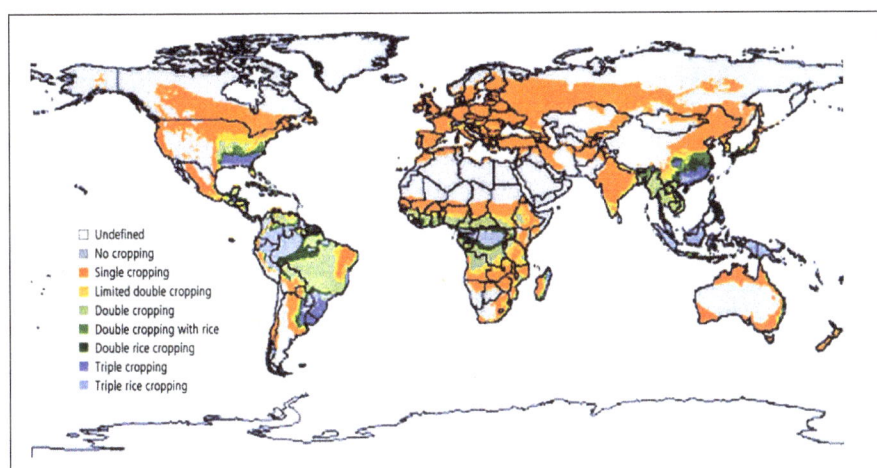

Multiple-cropping zones under rain-fed conditions

The potential area allowing double or triple cropping of rain-fed cereals is limited in the developed world. In Europe and Russia, it is virtually non-existent. In the developing world as a whole, about 55% of the land with the potential for rain-fed cultivation of cereals is suitable for double or triple cropping. In South and Central America it exceeds 80%: 65% for double and 17% for triple cropping.

Table: Share of cultivable land suitable for multiple cropping of rain-fed cereals at intermediate levels of inputs.

Region	Total suitable (million ha)	Single cropping (%)	Double cropping (%)	Triple cropping (%)
Oceania	87	83	13	4
Asia	366	63	32	5
Africa	678	59	39	2
Europe & Russia	422	100	<1	0
South & Central America	606	19	65	17

North America	297	57	30	13
Developing countries	1,645	45	47	8
Developed countries	811	82	13	5
World	2,455	57	36	7

Temperature and Rainfall Sensitivity

Global warming will lead to higher temperatures and changes in rainfall, and this in turn will modify the extent and productivity of land suitable for agriculture. The application of a set of temperature and rainfall sensitivity scenarios revealed a modest increase of cultivable rain-fed land for temperature increases up to 2 °C on a global scale. If temperature increases further but precipitation patterns and amounts remain at current levels, the extent of cultivable rain-fed land starts to decrease. When both temperature and rainfall amounts increase, the extent of cultivable rain-fed land increases steadily. For example, a temperature increase of 3 °C paired with a rainfall increase of 10% would lead globally to about 4% more cultivable rain-fed land. These figures are presented in table.

Table: Climate sensitivity of land suitable for cereal production (percentage change).

Region	Temperature increase			Temperature increase and rainfall change			
				+1°C,	+2°C,	+2°C,	+3°C,
	+1°C	+2°C	+3°C	+5%	+5%	+10%	+10%
Oceania	−4	−5	−9	1	0	4	0
Asia	4	3	−2	5	2	4	0
Africa	−4	−8	−12	−4	−8	−8	−13
Europe & Russia	13	17	20	16	21	24	28
Central America	−1	−4	−9	−4	−6	−9	−13
South America	−4	−11	−20	−7	−13	−14	−23
North America	12	16	20	16	21	24	28
Developing countries	−1	−6	−11	−2	−6	−6	−11
Developed countries	11	15	18	15	19	22	25
World	4	3	1	5	5	6	4

In the developed countries this increase is markedly higher, exceeding 25%. In contrast, the developing countries would experience a decrease of 11%, which could have serious consequences for food security in a number of poor developing countries in Africa, Asia, and South and Central America.

Limitations of the Approach

Important caveats concerning the Global AEZ results must be considered. While the study is based on the most recent global data compilations, the quality and reliability of these data sets are known to be uneven across regions, including, for example, soil data based on the FAO/UNESCO Soil Map of the World. Substantial improvements are being made to the soil information, and several regional updates have recently become available.

The current status of land degradation cannot be inferred from the Soil Map of the World. The only study with global coverage—the GLASOD or Global Assessment of Soil Degradation study —indicates that the status of land degradation may very well affect potential productivity of land. However, this degradation study offers insufficient detail and quantification for application within the Global AEZ model.

Socioeconomic needs of rapidly increasing and wealthier populations are the main driving forces in allocating land resources to various uses. Such considerations are critical for rational planning of sustainable agricultural development. As an extension of basic land productivity assessments, IIASA and the FAO have introduced interactive multiple-criteria model analysis for use in national and sub-national resource planning. It is at this level of analysis that socioeconomic considerations can be effectively taken into account.

Though various modes have been pursued for ground-truth control and verifying results of the Global AEZ suitability analysis, there is a need for further validation of the results and underlying databases.

Agroecological Zones and the Assessment of Crop Production Potential

The steady growth in the world's grain harvest from 1950 to 1990 has stalled during the Nineties. The 1995 world grain harvest of 1680 million tonnes (Mt), was the smallest in the previous seven years. The growing population puts considerable pressure on the scarce natural resources, and today it is well recognized that future productivity increases will have to be achieved while at the same time conserving and enhancing the natural resource base on which they depend.

The 'food first' focus of agricultural productivity research in the past focused primarily on 'vertical' relationships, i.e. investigations on plants, animals, air, water and soils within a relatively homogeneous spatial unit. However, the present task of conserving and enhancing natural resources is more complex and carries an implicit recognition of the presence of several systems within a landscape, which calls for 'horizontal analyses' requiring investigations and planning at temporal/spatial scales greater than the case with conventional studies. This requires integration of biological, physical and socio-economic factors in a holistic manner, and the availability of tools such as geographic information systems (GIS) and other spatial modelling techniques make this possible.

Agroecological Zones (AEZS) Approach

Agroecology is the application of ecological concepts and principles to the design and management of sustainable agricultural systems. The agroecological environment of a crop, land use or a farming system has physical, chemical and biological aspects. Viewing the agroecosystem as a functional system of complementary relations between living organisms and their environment that are managed by humans with the purpose of establishing agricultural production provides a basis for integrating the over- lapping ecological and environmental traits with sociological, economic, political, and other cultural components of agriculture. All of these may vary across space and time. Consequently, varieties and management methods have different optima in different places.

The site–season or site–year can be considered as the basic unit of the environment. At any given site, there are aspects that are invariant and stable, e.g. landform, soil type, etc., and aspects that vary with the seasons, e.g. temperature, rainfall and incidence of pests and diseases. Also, the severity or timing of aspects may fluctuate from year to year, depending on, for example, diseases, length of the growing season, etc. These give rise to a potentially infinite number of environments, but only part of the variation is of practical importance. Lack of rec- ognition of such site-specific characteristics, which can have a major influence on crop productivity, frequently led to disasters in the past. A good example is the case of groundnut cultivation in East Africa under un- suitable conditions during the 1950s, which resulted in huge financial losses. It is therefore important to summarize, classify and map environmental infor- mation at different levels of generalization. The resulting classes and maps range from very broad to very detailed AEZs.

Agroecological zonation helps mainly to (i) make a quantitative assessment of the biophysical resources upon which agriculture and forestry depend; and (ii) identify location-specific changes necessary to increase food production, through a comparison of farming systems and production alternatives. With AEZs, we can at least make a good estimate of the actual potential for crop production in different areas by (i) setting priorities for use of increased inputs that are needed to increase agricultural production, and (ii) identifying the less favoured environments, in which rural populations are ecologically and environmentally disadvantaged, and the priorities for their development.

Methods to Define AEZs

One of the simplest criteria for determining agroecological zonation has been mean rainfall. For example, in West Africa rainfall shows a significant north–south gradient because of the interseasonal movement of the Intertropical Convergence Zone (ITCZ) north and south of the equator. Hence, a range of vegetation patterns developed along this gradient. In almost all the climatic zonation schemes developed for West Africa, considerable emphasis has been placed on the use of two criteria: mean annual rainfall and veg- etation. These schemes are inadequate for assessing crop production potential because (i) mean annual rainfall cannot be considered as a sufficiently useful index of probable season length by itself as the potential evapotranspiration, which varies from one region to another, influences the proportion of rainfall available for crop growth and hence the final yields; (ii) for annual cereals which are planted and harvested fol- lowing rainfall patterns in a given year, the most important constraint is the length of the growing period (LGP); (iii) concepts and principles developed at a particular loca- tion in a given climatic zone, such as the Sahel, cannot be expected to hold good for the entire zone since soil characteristics such as texture, slope and water holding capacity and inherent soil fertility all play very important roles in cultivar performance; and (iv) the adoption of more acceptable, crop-dependent climatic criteria is necessary for crop planning applications.

The work of Cocheme & Franquin on the agroclimatological survey of a semi-arid area in Africa south of the Sahara should be considered a landmark in clarifying many issues of crop–climate relationships. They gave adequate weighting to both precipitations (P) and potential evapotranspiration (PE) in the zonation scheme by using the ratio of P: PE and computing the LGP based on a realistic appraisal of crop response to available moisture. This system was used by the UN Food and Agriculture Organization in their publication on agroclimatological data for Africa.

TAC/CGIAR Classification of Agroecological Zones

The Technical Advisory Committee (TAC) of the Consultative Group on Internation- al Agricultural Research (CGIAR) adapted the AEZ characterization developed by the FAO (TAC/CGIAR 1992) by subdividing the major ecological regions (tropics, subtrop- ics with summer or winter rainfall, and tem- perate regions) into rainfed moisture zones using LGP, and into thermal zones using the temperature regime prevailing during the growing period. Data on LGP, temperature, and the proportion of arable land covered by each of the nine zones are presented. Information on soil and landform used in the FAO characterization was excluded from the framework used by TAC so as to keep the number of subdivisions to a manageable level. Also, TAC's geographical coverage was limited to developing countries of sub- Saharan Africa, West Asia–North Africa, Asia, Latin America and the Caribbean. The current extent of cultivated land throughout the world is about 1.4 billion ha (of which 270 million ha are irrigated), but there is considerable variation in the percentage amounts of land used for arable cropping in the AEZs. According to TAC}CGIAR, in the nine AEZs in the developing countries of sub-Saharan Africa, Asia, West Asia–North Africa, Latin America and the Caribbean, about 950 million ha are cultivated and the total arable land is about 870 million ha. Land use patterns and the major food crops in different AEZs are shown in table.

Agro-climatic Suitability Classification

The agro-climatic suitability classification is based on:

i.   Matching the attributes of temperature regimes in each thermal zone to crop temperature requirements for photosynthesis and phenology, and determining whether the crop qualifies for further consideration in the matching exercise;

ii. Computation of constraint-free yield of the qualifying crop for each length of growing period zone;

iii. Computation of agronomically attainable yields for each crop in each length of growing period, taking into account yield losses that occur due to agro-climatic constraints of temperature and moisture stress, pests and diseases, and workability.

To enable crops to be matched to climatic conditions, the climatic inventory of Kenya was compiled to permit the interpretation of the climatic resources in terms of their suitability for production of crops. The appropriate climatic adaptability attributes of the crop dictate what parameters need to be taken into account in the compilation of the climatic inventory.

Crop Climatic Adaptability

Crops have climatic requirements for photosynthesis and phenology both of which bear a relationship to yield. The rate of crop photosynthesis and growth are related to the assimilation pathway and its response to temperature and radiation. However, the phenological climatic requirements, which must be met, are not specific to a photosynthetic pathway.

In the FAO Agro-ecological Zones methodology, crops are classified into climatic adaptability groups according to their fairly distinct photosynthesis characteristics. Each group comprises crops of 'similar ability' in relation to potential photosynthesis, and the differences between and within groups in the response of photosynthesis to temperature and radiation determine crop-specific biomass productivity when climatic phenological requirements are met.

Table: Average photosynthetic response of Individual leaves of four groups of crops to radiation and temperature.

Characteristics	Crop Adaptability Group			
	I	II	III	IV
Photosynthetic pathway	C_3	C_3	C4	C4
Rate of photosynthesis at light saturation at optimum temperature (mg $CO_2 dm^{-2}h^{-1}$)	20–30	40–50	> 70	> 70

Optimum temperature (°C)	15–20	25–30	30–35	20–30
Radiation intensity of maximum photosynthesis (cal cm^{-2} min^{-1})	0.2–0.6	0.3–0.8	>1.0	> 1.0
Crops included in the Kenya assessment	Barley Oat Wheat Phaseolus bean White potato	Cowpea Green gram Pigeon pea Phaseolus bean Rice Soybean Groundnut Sweet potato Cassava Banana Oil palm	Pearl millet Sorghum Maize Sugarcane	Sorghum Maize

Crop adaptability groups and their characteristic average photosynthesis response to temperature and radiation are presented in table. Barley and oat have a C3 photosynthesis pathway. They belong to group I and are adapted to operate under cool conditions (<20 °C mean daily temperature). Cowpea, green gram and pigeonpea have a C3 photosynthesis pathway. They belong to group II and are adapted to operate under warm conditions (>20 °C) with a potential rate of photosynthesis that is greater than in group I crops. Crops in group III (e.g. pearl millet, sugarcane) have a C4 photosynthesis pathway. They are adapted to operate under warm conditions (>20 °C) but with a potential rate of photosynthesis that is greater than in group II crops. Crops in group IV (e.g. highland maize) have a C4 photosynthesis pathway. They are adapted to operate under cool conditions (<20 °C) with a potential rate of photosynthesis that is similar to that in group III crops.

The time required to form yield depends on the phenological constraints on the use of time available in the growing period, and the location of yield in the plant (e.g. seed, leaf, stem, root) has an important influence. Temperature has a rate controlling/limiting effect on growth, and it may influence the growth of a specific part and the accumulation of yield if located therein. For example, in barley and oat, cool night temperatures are required for tillering but the optimum temperatures at the time of flowering and subsequent yield formation are higher. Similarly, optimum temperatures for growth in sugarcane are greater than 20 °C but during the ripening period, and because the yield is located in the stem, a lower temperature in the range 10–20 °C is required for concentration in the cane of sugar of the right kind. On the other hand, optimum temperatures for growth, development and yield formation in cowpea, green gram and pigeonpea are greater than 20 °C and most of the specific temperature requirements are also met when temperatures are optimum for photosynthesis and growth.

Table: Climatic adaptability attributes of crops.

Attributes	Barley	Oat	Cowpea	Green gram	Pigeon pea
Species	Hordeum vulgare	Avena sativa	Vigna unguiculata	Vigna radiata	Cajanus cajan
Photosynthetic pathway	C_3	C_3	C_3	C_3	C_3
Crop adaptability group	I	I	II	II	II
Days to maturity	90–120[1]	90–120[1]	80–100[4]	60–80[4]	30–150[4]
	120–150[2]	120–150[2]	100–140[4]	80–100[4]	150–170[4]
	150–180[3]	150–180[3]			170–190[4]
Harvested part	Seed	Seed	Seed	Seed	Seed
Main product	Grain (C)	Grain (C)	Grain (L)	Grain (L)	Grain (L)
Growth habit	Determinate	Determinate	Indeterminate	Indeterminate	Indeterminate
Life-span					
- Natural	Annual	Annual	Annual	Annual	Short-term perennial
- Cultivated	Annual	Annual	Annual	Annual	Annual/Biennial
Yield: Cultivated	TI	TI	LI	LI	LI
Formation period	LT	LT	ME	ME	ME
Thermal zone for consideration	3, 4, 5, 6, 7	3, 4, 5, 6, 7	1, 2, 3	1, 2, 3	1, 2, 3

C - Cereal, L - Legume, TI - Terminal inflorescence, LI - Lateral inflorescence, LT - Last one third of growth cycle, ME - Middle to end period of growth cycle

Thermal zones: 1 - <25.0 °C, 2 – 22.5–25.0, 3 – 20.0–22.5, 4 – 17.5–20.0, 5 – 15.0–17.5, 6 – 12.5–15.0, 7 – 10.0–12.5

1 thermal zones 3 & 4;   2 thermal zone 5;   3 thermal zones 6 & 7;
4 thermal zones 1,2 & 3

The attributes that are helpful in assessing the climatic adaptability of the crops in the matching exercise are given in table.

Barley and oat (C_3-species, group I) are annuals with a botanically determinate growth habit. Their yield is located in terminal inflorescences in seeds, and the crop yield formation period is the last one-third of their growth cycle. Their climatic adaptability attributes qualify them to be considered for matching in areas with mean daily temperatures less than 22.5 °C and more than 10°C (i.e. thermal zones 3, 4, 5, 6 and 7).

Cowpea (C_3-species, group II) is an annual with botanically indeterminate growth habit, offering cultivars that may be morphologically determinate (bunch types) or indeterminate (spreading types). Its yield is located in the lateral inflorescences in seeds, and the crop yield formation period is from the middle to the end of its growth cycle.

Its climatic adaptability attributes qualify it to be considered for matching in areas with mean daily temperatures greater than 20°C (i.e. thermal zones 1, 2 and 3).

Table: Thermal zones.

Thermal zone code	Temperature class (°C)	Altitude (m)	Crop group suitable for consideration			
1	< 25.0	> 800	II	III		
2	22.5 – 25.0	800–1200	II	III		
3	20.0 – 22.5	1200 – 1550	I	II	III	
4	17.5–20.0	1550–1950	I	II	III	IV
5	15.0 – 17.5	1950–2350	I	IV		
6	12.5 – 15.0	2350 – 2700	I	IV		
7	10.0 – 12.5	2700–3100	I			
8	5.0 – 10.0	3100–3900	I			
9	< 5.0	> 3900	I			

Green gram (C3-species, group II) is an annual with botanically indeterminate growth habit, offering cultivars that may be morphologically determinate in growth and stature. Its yield is located in the lateral inflorescences in seeds, and the crop yield formation period is from the middle to the end of its growth cycle. Its climatic adaptability attributes qualify it to be considered for matching in areas with mean daily temperatures greater than 20 °C (i.e. thermal zones 1, 2 and 3).

Pigeonpea (C3-species, group II) is a short-term perennial with botanically indeterminate but morphologically determinate growth habit. Its yield is located in the lateral inflorescences in seeds, and the crop yield formation period is from the middle to the end of its annual cultivated life-span. Its climatic adaptability attributes qualify it to be considered for matching in areas with mean daily temperatures greater than 20 °C (i.e. thermal zones 1, 2 and 3).

Climatic Inventory

Quantification of heat attributes has been achieved by defining reference thermal zones representing the prevailing temperature regimes. Temperature seasonality effects of latitude are minor due to the equatorial position of Kenya.

The temperature threshold used in these definitions accord with those differentiating the four temperature adaptability groups of crops.

Quantification of moisture conditions was achieved through the concept of reference length of growing period (LGP) being defined as the duration (in days) when moisture supply can permit crop growth. A moisture supply from rainfall of half, or more than half, potential evapotranspiration has been considered to permit crop growth. The following main concepts, definitions and methods form the basis of the quantification of moisture conditions in the climatic inventory.

The growing period is the time when moisture supply from rainfall exceeds half potential evapotranspiration. It includes the time required to evapotranspire up to 100 mm of stored moisture from the soil profile. A 'normal' growing period has a humid phase, i.e. a period when moisture supply is greater than full potential evapotranspiration. When there is no humid period, the growing period is defined as 'intermediate'.

The quantification of moisture regime is based on the analysis of the length of growing period for each year separately and the computation of:

a. Number of separate lengths of growing periods per year, summerized as a historical profile of pattern of number of growing periods per year (referred to as LGP-Pattern);

b. Length of each growing period and its various moisture periods,summarized as mean total dominant length, first associated length and second associated length, and the mean individual dominant and associated lengths making up the total lengths;

c. The quality of moisture conditions during the growing period and its various moisture periods;

d. Year-to-year-variability (frequency distribution) of each length of growing period and the associated moisture condition.

Twenty two LGP-Patterns are recognized, and these with their composition are presented in table. The LGP-Pattern code represents the number of growing periods per year in order of frequency of occurrence, e.g. in the pattern coded 2-1-3, the numeral 2 represents the number of lengths of growing periods per year (i.e. two)that occur in the majority of the years (i.e. 55 percent) - the dominant length number; the number 1 represents number of lengths of growing periods per year (i.e. one) that has the next most commonly occurring frequency (i.e. 25 percent) the first associated length number; and the numeral 3 represents number of lengths of growing periods per year (i.e. three) that has the smallest occurrence (i.e. 20 percent) - the second associated length number.

Table: Patterns of growing periods (LGP-patterns) - historical profiles of occurrence of number of growing periods per year.

Code	LGP-Pattern	Proportion (%)
1	1	100
2	H - 1	60 : 40
3	1 - H	70 : 30
4	1 - H - 2	65 : 20 : 15
5	1 - 2 - H	65 : 20 : 15
6	1 - 2	65 : 35
7	1 - 2 - 3	50 : 35: 15
8	1 - 3 - 2	50 : 30 : 20

9	1 - 2 - D	40 : 35 : 25
10	1 - D - 2	40 : 35 : 25
11	1 - D	60 : 40
12	2	100
13	2 - 1	70 : 30
14	2 - 1 - H	55 : 30 : 15
15	2 - 1 - 3	55 : 25 : 20
16	2 - 3	75 : 25
17	2 - 3 - 1	60 : 25 : 15
18	2 - 3 - 4	60 : 30 : 10
19	2 - 1 -D	70 : 15 : 15
20	3 - 2	60 : 40
21	3 - 2 - 1	50 : 35 : 15
22	D	100

H = 365 + days (i.e. year-round humid), D = zero days (i.e. year-round dry).

For each LGP-Pattern type, the mean total length of the dominant number is correlated with the mean total length of the associated numbers. Also, when the mean total length is a summation of more than one mean length, the latter lengths are again correlated to the former total length. These relationships are presented in tables.

In the climatic inventory of Kenya, only the mean total dominant length has been inventoried on the map as 14 LGP zones. The boundary or isoline values used are 0, 30, 60, 90, 120, 150, 180, 210, 240, 270, 300, 330, 365 and 365+ days respectively delineating the mean total dominant length of growing period zones of 0, 1–29, 30–59, 60–89, 90–119, 120–149, 150–179, 180–209, 210–239, 240269, 270–299, 300–329, 330–364, 365- and 365+days.

Additionally, the LGP-Pattern zones have been inventoried. Consequently, the relationships in tables below together with the map of dominant LGP zones and the LGP-Pattern zones provide the historical profile of any mean total dominant length of growing period in any of the 22 LGP-Pattern zones.

Table: Relationships between mean total dominant and mean total associated lengths of growing period.

LGP-Pattern	Relationship
1 - 2	$L2 = 80.40 + 0.75\ L1$
1 - 2 - H	
1 - H - 2	
1 - 2 - 3	$L2 = 71.56 + 0.77\ L1$
1 - 3 - 2	$L3 = 77.14 + 0.66\ L1$
1 - 2 - D	

1 - 2 - D	
2 - 1	$L1 = -86.09 + 1.28 \, L2$
2 - 1 - H	$L3 = 25.29 + 0.82 \, L2$
2 - 1 - 3	
2 - 1 - D	
2 - 3	$L3 = 30.11 + 0.83 \, L2$
2 - 3 - 1	$L1 = -98.72 + 1.35 \, L2$
2 - 3 - 4	$L4 = 114.54 + 0.58 \, L2$
3 - 2	$L2 = 45.05 + 0.80 \, L3$
3 - 2 - 1	$L1 = -9.86 + 0.88 \, L3$

L1 = Total length of one growing period per year, L2 = Total length of two growing periods per year,

L3 = Total length of three growing periods per year, L4 = Total length of four growing periods per year.

Matching of Crops to Thermal Zones

The initial step in the matching process is comparison of the temperature requirements of individual crops with the identified thermal zones. This step indicates the crops which should be considered from a temperature/growth and phenology viewpoint, in each thermal zone.

Crop/thermal zone suitability ratings for each crop and zone are presented in table. Five suitability classes are employed (i.e. S1, S2, S3, S4, and N), and the ratings apply to both levels of inputs: where requirements are fully met, the zone is adjusted S1: where requirements are sub-optimal, the zone is adjudged S2, S3 or S4; where requirements are not met, the zone is adjudged as N.

A rating of S1 indicates that the temperature conditions for growth and yield physiology, and phenology development are optimal and that it is possible to achieve the maximum attainable agro-genetic yield potential,if there are no additional climatic or edaphic limitations. Ratings of S2, S3 and S4 indicate that temperature conditions for growth and development are sub-optimal and that there would be a suppression of yield potential in the order of 25, 50 and 75 percent respectively. A rating of N indicates that the thermal requirements are not met and the zone is not suitable for father consideration.

Table: Relationship between individual component mean length and mean total length of growing period.

LGP-Pattern	Relationship
2	$L2_1 = -1.11 + 0.55 \, L2$
1 - 2	$L2_1 = 4.94 + 0.62 \, L2$
1 - 2 - H	
1 - H - 2	

1 - 2 - 3	$L2_1 = 5.87 + 0.64\,L2$
1 - 3 - 2	$L3_1 = 22.12 + 0.39\,L3$
1 - 2 - D	$L3_2 = 1.58 + 0.32\,L3$
1 - D - 2	
2 - 1	$L2_1 = -5.48 + 0.64\,L2$
2 - 1 - H	$L3_1 = 0.14 + 0.46\,L3$
2 - 1 - 3	$L3_2 = -0.98 + 0.33\,L3$
2 - 1 - D	
2 - 3	$L2_1 = -3.05 + 0.61\,L2$
2 - 3 - 1	$L3_1 = 1.68 + 0.43\,L3$
2 - 3 - 4	$L3_2 = -3.00 + 0.34\,L3$
	$L4_1 = 26.35 + 0.34\,L4$
	$L4_2 = -20.88 + 0.38\,L4$
	$L4_3 = -17.66 + 0.27\,L4$
3 - 2	$L2_1 = -2.33 + 0.63\,L2$
3 - 2 - 1	$L3_1 = 5.62 + 0.45\,L3$
	$L3_2 = 1.25 + 0.31\,L3$

$L2_1$ = First length of the two growing periods per year,

$L3_1$ = First length of the three growing periods per year,

$L3_2$ = Second length of the three growing periods per year,

$L4_1$ = First length of the four growing periods per year,

$L4_2$ = Second length of the four growing periods per year,

$L4_3$ = Third length of the four growing periods per year.

Matching of Crops of Growing Period Zones

Matching of crops of growing period zones is according to the following procedure:

i.   Assessment of net biomass and constraint-free crop yields by individual lengths of growing period zones, assuming optimum temperature conditions for production (i.e. S1 crop/thermal zone rating);

ii. Inventory of agro-climatic constraints for each length of growing period zone by crop and by input level;

iii. Application of the thermal zone suitability ratings and agro-climatic constraints (ii) to the constraint-free yields (i) to determine (agro-climatically attainable) crop yields by individual lengths of growing period zones in each thermal zone.

Table: Thermal zones suitability ratings

Crop	Growth cycle (days)	Thermal zones								
		T1	T2	T3	T4	T5	T6	T7	T8	T9
Barley	90–120	N	N	S3	S1	na	na	na	N	N
Barley	120–150	N	N	na	na	S1	na	na	N	N
Barley	150–180	N	N	na	na	na	S2	S4	N	N
Oat	90–120	N	N	S4	S2	na	na	na	N	N
Oat	120–150	N	N	na	na	S1	na	na	N	N
Oat	150–180	N	N	na	na	na	S2	S4	N	N
Cowpea	80–100	S1	S1	S3	N	N	N	N	N	N
Cowpea	100–140	S1	S1	S3	N	N	N	N	N	N
Green gram	60–80	S1	S2	S4	N	N	N	N	N	N
Green gram	80–100	S1	S2	S4	N	N	N	N	N	N
Pigeonpea	130–150	S1	S1	S3	N	N	N	N	N	N
Pigeonpea	150–170	S1	S1	S3	N	N	N	N	N	N
Pigeonpea	170–190	S1	S1	S3	N	N	N	N	N	N

The above matching exercise results in a basic agro-climatic suitability classification of each length of growing period zone by thermal zone only. From this, the agro-climatic suitability classification of each mean total dominant growing period zone (inventoried) can be derived for each crop according to agro-climatically attainable yields by thermal zone and by pattern zone. This is achieved by computing agro-climatically attainable yields as affected by year-to-year variability (i.e. LGP-Pattern) from the basic agro-climatic suitability classification of each length of growing period.

Potential Net Biomass and Yield

The methodology for the calculation of net biomass and constraint-free yields by suitable thermal zone is according to Kassam and is presented below.

Net total biomass (Bn) is calculated from the equation:

$$Bn = (0.36 \text{ bgm} \times L) / (1/N + 0.25 \text{ Ct})$$

Where, Bgm = maximum rate of gross biomass production at leaf area index (LAI) of 5.

L = maximum growth ratio, equal to the ratio of bgm at actual LAI to bgm at LAI of 5. (L at LAI 1, 2, 3, 4 and 5 is 0.4, 0.6, 0.8, 0.9 and 1.0 respectively).

N = length of crop growth cycle.

Ct = maintenance respiration, dependent on both crop and temperature; given by the relation:

$$Ct = C30 (0.0044 + 0.0019\ T + 0.0010\ T^2)$$

At 30 °C, C = 0.0283 for a legume crop and 0.0108 for a non-legume crop.

Constraint-free yield (By) is calculated from net biomass (Bn) from the equation:

$$By = Hi \times Bn$$

Where, Hi = Harvest index (i.e. proportion of the net biomass of the crop that is economically useful).

The maximum rate of gross biomass production (bgm) is dependent on the maximum rate photosynthesis (Pm) which is dependent on temperature and photosynthesis pathway of the crop. Maximum rates of photosynthesis (Pm) for the five crops by temperature is presented in table.

For Pm = 20 kg CH_2O ha^{-1} hr^{-1} and LAI of 5, bgm is calculated from the equation:

$$bgm = F \times bo + (1\text{-}F)\ bc$$

Where, F = fraction of the daytime the sky is clouded:

$$F = (Ac - 0.5\ Rg)/(0.8\ Ac)$$

Where, Ac is the maximum active incoming shortwave radiation on clear days in cal cm^{-2} day^{-1} and Rg is the incoming shortwave radiation in cal cm^2 day^{-1}.

bo = gross dry matter production rate of a standard crop for a given location on a completely overcast day, kg CH_{20} ha^{-1} day^{-1}.

bc = gross dry matter production rate of standard crop for a given location on a clear (cloudless) day, kg CH_{20} ha-1 day^{-1}

When Pm is greater than 20 kg CH_{20} ha^{-1} hr^{-1}, bgm is given by the equation:

$$bgm = F(0.8 + 0.01\ Pm)bo + (1 - F)(0.5 + 0.025Pm)bc.$$

When Pm is less than 20 kg CH_{20} ha^{-1} hr^{-1}, bgm is given by the equation:

$$bgm = F(0.5 + 0.025Pm)bo + (1 - F)(0.05Pm)bc.$$

Table: Maximum rate of photosynthesis. (Pm in kg CH_{20} ha^{-1} hr^{-1})

Crops	Average day-time temperature (°C)				
	10	15	20	25	30
Barley and oat	15	20	20	15	5
Cowpea, green gram and pigeonpea	0	15	32.5	35	35

Agro-edaphic Suitability Classification

As a medium in which roots grow and as a reservoir for water and nutrients on which crops continuously draw during their life cycle, soils are natural resource and valuable economic asset requiring protection, conservation and improvement through good husbandry.

The adequate agricultural exploitation of the climatic potential and sustained maintenance of productivity largely depends on soil fertility and management of soil on an ecologically sound basis. Soil fertility is concerned with the ability of the soil to supply nutrients and water to enable crops to maximize the climatic resources of a given location. The fertility of a soil is determined by its both physical and chemical properties whose understanding is essential to the effective utilization of climate and crop resources for optimum production.

In order to assess suitability of soils for crop production, soil requirements of crops must be known. Further, these requirements must be understood within the context of limitations imposed by landform and other features which do not form a part of soil but may have a significant influence on the use that can be made of the soil.

The basic soil requirements of crop plants may be summarized under the following headings, related to internal and external soil properties:

a. Internal requirements:

  • The soil temperature regime, as a function of the heat balance of soils as related to annual or seasonal and/or daily temperature fluctuations;

  • The soil moisture regime, as a function of the water balance of soils as related to the soil's capacity to store, retain, transport and release moisture for crop growth, and/or to the soil's permeability and drainage characteristics;

  • The soil aeration regime, as a function of the soil air balance as related to its capacity to supply and transport oxygen to the root zone and to remove carbon dioxide;

  • The natural soil fertility regime, as related to the soil's capacity to store, retain and release plant nutrients in such kinds and proportions as required by crops during growth;

  • The effective soil depth available for root development and foothold of the crop;

Table: Crop edaphic adaptability inventory

Crop	Slope (Percent)				Drainage	
	High inputs		Low & Int. inputs		All inputs	
	Optimum	Marginal	Optimum	Marginal	Optimum	Range
Barley	0–8	8–16	0–8	8–24	MW-W	I-SE
Oat	0–8	8–16	0–8	8–24	MW-W	I-SE
Cowpea	0–8	8–16	0–8	8–20	MW-W	I-SE
Green gram	0–8	8–16	0–8	8–20	MW-W	I-SE
Pigeon pea	0–8	8–16	0–8	8–20	MW-W	I-SE

Drainage classes: I = imperfectly drained; MW = moderately well drained; W = well drained; SE = somewhat excessively drained; E = excessively drained.

Crop	Flooding		Texture			
	All Inputs		High inputs		Low & Int. inputs	
	Optimum	Marginal	Optimum	Range	Optimum	Range
Barley	F_0	F_1	L-MCs	SL-MCs	L-SC	SL-KC
Oat	F_0	F_1	L-C	SL-MCs	L-SC	SL-KC
Cowpea	F_0	F_1	SL-SCL	LS-KC	SL-SCL	LS-KC
Green gram	F_0	F_1	L-CL	SL-KC	L-CL	LS-KC
Pigeon pea	F_0	F_1	SL-SCL	LS-KC	SL-SCL	LS-KC

Flooding classes: F_0 = no floods; F_1 = occasional flooding.

Texture classes: MCs = montmorillonitic clay, structured; C = clay (mixed unspecified); KC = kaolinitic clay; SC = sandy clay; SiCL = silty clay loam; CL = clay loam; SCL = sandy clay loam; L = loam; SL = sandy loam; LS = loamy sand.

CROP	DEPTH (cm)		$CaCO_3$(%)		GYPSUM (%)	
	All inputs		All inputs		All inputs	
	Optimum	Marginal	Optimum	Marginal	Optimum	Marginal
Barley	> 50	25–50	0–30	30–60	0–5	5–20
Oat	> 50	25–50	0–30	30–60	0–5	5–20
Cowpea	> 75	50–75	0–20	20–35	0–3	3–15
Green gram	> 75	50–75	0–25	20–35	0–3	3–15
Pigeon pea	> 100	50–100	0–25	20–50	0–3	3–15

Crop	PH		Fertility requirements	Salinity (mmhos/cm)	
	All inputs		All inputs	All inputs	
	Optimum	Range	Range	Optimum	Range
Barley	6.0–7.5	5.2–8.5	moderate	0–8	8–12

Oat	6.0–7.5	5.2–8.2	low/moderate	0–5	5–10
Cowpea	5.2–7.5	5.0–8.2	low/moderate	0–3	3–6
Green gram	5.5–7.5	5.2–8.2	moderate	0–3	3–6
Pigeon pea	5.2–7.5	5.0–8.2	low/moderate	0–3	3–6

| Crop | Alkalinity (ESP) | |
| | All inputs | |
	Optimum	Marginal
Barley	0–35	35–50
Oat	0–30	30–45
Cowpea	0–5	8–12
Green gram	0–5	8–12
Pigeon pea	0–5	8–12

- Soil texture at the surface and within the whole depth of soil required for normal crop development;

- The absence of soil salinity and of specific toxic substance or ions deleterious to crop growth;

- Other specific properties, e.g. soil tilth as required for germination and early growth.

b. External requirements: in addition to the above internal soil requirements of crops, a number of external soil requirements are of importance, e.g.:

- Soil slope, topography and characteristics determined by micro and macrorelief of the soil;

- Occurrence of flooding as related to crop susceptibility to flooding during the growing period;

- Soil accessibility and trafficability under certain management systems.

Crop Edaphic Adaptability

From the basic soil requirements of crops, a number of crop response related soil characteristics can be derived. One of these characteristics is, for instance, soil pH. For most crops and cultivars, optimal soil pH is known and can be quantified by a range within which it is not limiting to growth. Outside the optimal range, there is a critical range within which the crop can be grown successfully but with diminished yield. Beyond the critical range, the crop cannot be expected to yield satisfactorily unless special precautionary management measures are taken.

The same holds for other soil requirements of plants related to soil characteristics. Many soil characteristics can be defined in a range that is optimal for a given crop, a range that is critical or marginal, and a range that is unsuitable under present technology.

Table presents for barley, oat, cowpea, green gram and pigeonpea, optimal and critical ranges of the following soil characteristics: soil slope, soil depth, soil drainage, flooding, texture and clay type, natural fertility (including cation exchange capacity, percent base saturation and organic matter), salinity, pH, free calcium carbonate content and gypsum content.

Many of the soil characteristics are at least partly intrinsically related to the soil. This relationship has guided the definition of optimal and marginal ranges of the various soil characteristics and so simplified the subsequent matching of the different soil units with the inventoried soil requirements of crops.

Soil Inventory

The soil resources of Kenya have been inventoried in terms of associations of soil units, and the corresponding characterization of soil textures, phases, stoniness and slopes.

Soil units have been defined in terms of measurable and observable properties of the soil itself, and specific clusters of such properties are combined into 'diagnostic horizons' and soil units.

Soil texture may vary within the range of textures defined for a particular soil unit. In the legend of the Exploratory Soil Map, textural classes for the individual soil units by soil mapping unit are presented. The three major textural divisions (coarse, medium and fine) are subdivided into 17 classes Soil phases indicate land characteristics which are not considered in the definition of the soil units but are significant to the use and management of land. Soil phases recognized on the Exploratory Soil Map of Kenya can be grouped into phases indicating a mechanical hindrance or limitation (rocky, bouldery, boulder-mantle, stony, stone-mantle, gravel-mantle), phases indicating an effective soil depth limitation (lithic, paralithic, petro-calcic, piso-calcic, petro-ferric, piso-ferric), and phases indicating a physico-chemical limitation (saline, sodic and saline-sodic). Soil phases occur either individually or in combinations of up to three.

Table: Texture.

Texture Symbol	Texture class
Coarse:	
S	Sand
LCS	Loamy coarse sand
FS	Fine sand
LFS	Loamy fine sand
LS	Loamy sand LS

Medium:	
FSL	Fine sandy loam
SL	Sandy loam
L	Loam
SCL	Sandy clay loam
SL	Silt loam
CL	Clay loam
SIL	Silty clay loam
SI	Silt
Fine:	
SC	Sandy clay
SIC	Silty clay
PC	Peaty clay
C	Clay

The presence of coarse material (stoniness) in the soil profile has been inventoried seperately from soil textures. Six types of coarse material or stoniness have been inventoried: Gravelly (G), Very Gravelly (VG), Stony (S), Bouldery (B), Stony/Bouldery (SB) and Bouldery/Stony (BS).

Six basic slope classes, in 12 combinations, have been employed in the Exploratory Soil Map of Kenya. The six basic slope classes are: A: 0–2%; B: 2–5%; C: 5–8%; D: 8–16%; E: 16–30% and F: > 30%. The 12 combination slope classes are: A: 0–2%; AB: 0–5%; B: 25%; BC: 2–8%; C: 5–8%; BCD: 2–16%; CD: 5–16%; D: 8–16%; DE: 8–30%; E: 16–30%; EF: 16->30%; F: >30%.

To each of these 12 slope classes, associated slope classes have been assigned. These associated slope classes, covering upto 10% of the land area of the 12 slope classes, are used for evaluation purposes only. They are not included explicitly in the soil resources inventory. The 12 inventoried combination slope classes and the associated slope classes are presented in table. For the same purposes of evaluation, assumed mean slopes have been assigned to each of the quartiles of the land area of each of the 12 slope classes.

Table: Soil phase.

Symbol	Name	Symbol	Name	Symbol	Name
Single.:		Combination of two:		Combination of three:	
R	Rocky	R/B	Rocky and bouldery	R/B/AO	Rocky and bouldery and saline-sodic
B	Bouldery	R/S	Rocky and Many	ROTS	Rocky and lithic and stony
BM	Boulder-mentle	B/S	Bouldery and stony	B/S/A	Bouldery and stony and saline

S	Stony	BM/AO	Boulder-mantle and saline-saline-sodic	BM/S/AO	Bouldery and atony and saline-sodic
SM	Stone mantle	S/R	Stoney and rocky	P/R/B	Lithic and rocky and bouldery
G	Gravelly	S/B	Stony and bouldery	P/R/S	Lithic and rocky and atony
GM	Gravel-mantle	S/K	Stony and pertrocalcic	P/B/S	Lithic and bouldery and atony
P	Lithic	S/AO	Stony and saline-sodic	P/B/A	Lithic and bouldery and saline
PP	Paralithic	SM/O	Stone mantle and sodic	P/BM/AO	Lithic and bouldery-mantle and saline-sodic
K	Petrocalcic	SM/AO	Stone mantle and saline-sodic	P/S/R	Lithic and atony and rocky
KK	Petrocalcic	P/R	Lithic and rocky	P/S/A	Lithic and atony and saline
C	Pisocalcic	P/B	lithic and bouldery	P/S/AO	Lithic and atony and saline-sodic
CC	Pisocalcic	P/BM	Lithic and boulder-mantla	P/SM/AO	Lithic and stone-mantle and saline-sodic
M	Petroferric	P/S	Lithic and stony	P/GM/S	Lithic and gravel-mantle and saline
MM	Pisoferric	P/O	Lithic and sodic		
A	Saline	P/AO	Lithic and saline-sodic		
O	Sodic	PP/R	Paralithic and rocky		
AO	saline-sodic	PP/S	Paralithic and atony		
F	Fragipan	K/S	Petrocalcic and stony		
		K/A	Petrocalcic and saline-sodic		
		KK/A	Petrocalcic and saline		
		KK/O	Petrocalcic and sodic		
		M/R	Pisoferric and rocky		
		M/M	Pisoferric and pisoferric		
		A/F	Pisoferric and fragipan		
		O/F	Sodic and fragipan		

Agro-edaphic Suitability

From the basic soil requirements of crops, a number of responses related soil charac-teristics can be derived. A correlation between soil requirements listed above and soil characteristics that can be used as soil factors to rate crop performance is given in table.

As explained earlier, the soil units have been defined in terms of measurable and observable properties of the soil itself, and specific clusters of such properties are combined into 'diagnostic properties', which are used in the definition of the soil units.

Table: Associated slope classes.

Slope class		Associated slope classes							
symbol	%								
A	0 – 2	100%	A						
AB	0 – 5	100%	AB						
B	2 – 5	100%	B						
BC	2 – 8	90%	BC	5%	A	5%	D		
C	5 – 8	90%	C	5%	AB	5%	D		
BCD	2 – 16	90%	BCD	5%	A	5%	E		
CD	5 – 16	90%	CD	5%	AB	5%	E		
D	8 – 16	90%	D	5%	BC	5%	E		
DE	8 – 30	90%	DE	5%	BC	5%	F		
E	16 – 30	90%	E	5%	BCD	5%	F		
EF	16 – 56	95%	EF	5%	BCD				
F	30 – 56	95%	F	5%	DE				

The diagnostic horizons have been used as defined in the FAO/Unesco Soil Map of the World Legend. Diagnostic properties however have been narrowed down in case of ferric properties and widened in case of vertic properties.

Table: Quartiles of slope classes.

Slope	class	Gentlest	Lower	Upper	Steepest
symbol	%	Q1	Q2	Q3	Q4
A	0 – 2	0	1	1	2
AB	0 – 5	0	2	4	5
B	2 – 5	2	3	4	5
BC	2 – 8	2	4	6	8
C	5 – 8	5	6	7	8
BCD	2 – 16	2	6	11	16
CD	5 – 16	5	9	12	16
D	8 – 16	8	11	13	16
DE	8 – 30	8	16	22	30
E	16 – 30	16	21	25	30
EF	16 – 56	16	30	42	56
F	30 – 56	30	39	47	56

Table: Relation between basic soil requirements for crops and soil characteristics.

Basic soil requirements	Soil characteristics (soil factors)
Moisture availaibility	- Effective soil depth
	- Available soil moisture holding capacity
	- Drainage
Nutrient availability	- Nutrient availability
	- Soil reaction
Oxygen availability	- Soil permeability
	- Drainage
Foothold for roots	- Effective soil depth
Salinity	- Soil salinity
Toxicity	- Soil reaction
Accessibility and trafficability (workability)	- Topsoil consistency and bearing capacity
Soil tilth for crop establishment	- Topsoil consistency and bearing capacity

Histic H horizon	Surface layer of organic material more than 20 cm thick.
MOLLIC A horizon	Surface horizon with dark colour, medium to high humus content, high base saturation.
Umbric A horizon	Surface horizon with dark colour, medium to high humus content, low base saturation.
Ochric A horizon	Surface horizon with light colour, low humus content.
Argillic B horizon	Subsoil horizon with accumulation of illuvial clay.
Natric B horizon	Subsoil horizon with accumulation of illuvial clay and high exchangeable sodium.
Cambic B horizon	Subsoil horizon with a structure and/or colour different from overlying and underlying horizons.
Spodic B horizon	Subsoil horizon with accumulation of iron and/or humus.
Oxic B horizon	Subsoil with residual accumulation of sesquioxides and low CEC.
Calcic horizon	Horizon of accumulation of calcium carbonate.
Gypsic horizon	Horizon of accumulation of calcium sulphate.
Sulphuric horizon	Horizon with strong acidity and prominent mottling.
Albic E horizon	Eluvial horizon from which clay and free iron oxide have been removed, light colour.
Calcareous material	Calcium carbonate present at least between 20 and SO cm from the surface. CEC high or very high: Exchange complex dominated by allophane or montmorillonite.
CEC low	Exchange complex dominated by kaolinite (CEC less than 24 meq/100 g clay).
CEC very -low	Less than l.5 meq/100 g clay.
Cracking clays	Formation of deep and wide cracks upon drying.
Plinthite	Mottled subsoil layer which irreversibly hardens upon exposure to repeated wetting and drying.

High salinity	Electrical conductivity (EC) higher than 15 mmhos/cm.
Moderate salinity	Electrical conductivity (EC) between 4 and 15 mmhos/cm.
High alkalinity	Saturation with exchangeable sodium of more than 15 percent.
Moderate alkalinity	Saturation with exchangeable sodium of 6 to 15 percent.
Indurated subsoil	Subsoil layer with firm or hard consistence, but can still be penetrated by spade or auger.
Cemented hardpan	Extremely hard continuous subsoil layer which cannot be penetrated by spade or auger.
Coarse texture	Less than 18 percent clay and more than 65 percent sand.
Heavy texture	More than 35 percent clay.
Abrupt textural change	Considerable increase in clay content within a very short vertical distance.
Tonguing	Deep and irregular penetration of an albic E horizon into an argillic B horizon

The agro-edaphic suitability classification is input-specific and based on:

i. Matching the soil requirements of crops with the soil conditions of the soil units described in the soil inventory (soil unit evaluation), and

ii. Modification of the soil unit evaluation by limitation imposed by, texture, stoniness, phase and slope conditions.

Soil Unit Evaluation

The soil unit evaluation is expressed in terms of ratings based on how far the soil conditions of a soil unit meet crop requirements under three specified levels of inputs. The appraisal is effected in five basic classes for each crop and level of inputs, i.e. very suitable (S1), suitable (S2), moderately suitable (S3), marginally suitable (S4), and not suitable (N).

A rating of S1 indicates that the soil conditions for crop production are optimal and that suppression of potential yields (if any) is assumed to be slight or nil. The rating of S2 indicates that there are slight to moderate limitation which would suppress potential yields by some 25 percent. The rating of S3 indicates sub-optimal soil conditions with moderate to severe limitations which would suppress potential yields by some SO percent. The rating of S4 indicates sub-optimal soil conditions with severe limitations which would suppress potential yields by some 75 percent. The rating of N indicates that crop production is not possible.

Texture Evaluation

All ratings of soils with coarse texture (sand, loamy coarse sand, fine sand, loamy fine sand and loamy sand) for the five crops are classified one class lower for 50% and two classes lower for the remaining 50% of its extents, except for Arenosols (Q), Albic Arenosols (Qa), Cambic Arenosols (Qc), Ferralic Arenosols (Qf), Calcaro-cambic

Arenosols (Qkc), Luvic Arenosols (Ql) and Vitric Andosols (Tv), which should remain unchanged since coarse texture limitations have already been applied in the soil unit ratings. All ratings of soils with medium and fine textures remain unchanged since limitations imposed by these textures have been included in the soil unit ratings.

Stoniness Evaluation

The limitations imposed by presence of coarse material (stoniness) in the soil profile have been rated (using the five basic classes described in) by crop and inputs level.

Soil Phase Evaluation

The limitations imposed by presence of soil phases which occur individually or in combinations of up to three phases have been rated (using the five basic classes described above) by crop and inputs level.

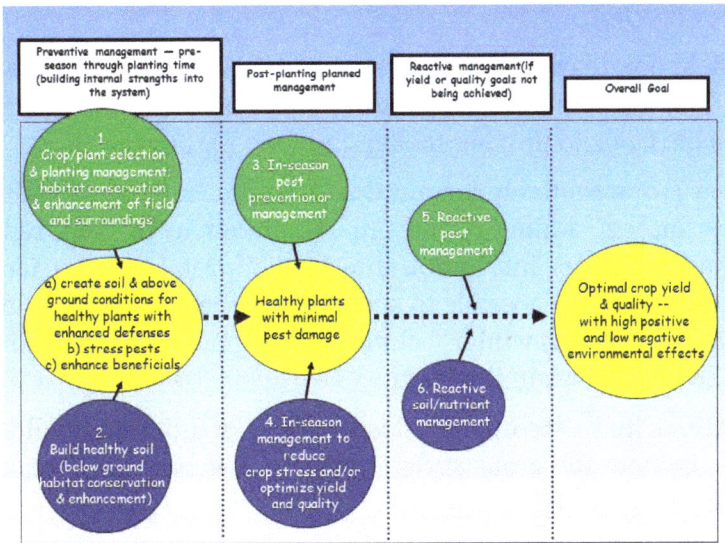

A whole-system approach to soil and crop management at the field level.

The heart of the matter is that the strength of the system is improved by creating improved habitat both above ground and in the soil. Although it is somewhat artificial to talk separately about aboveground and soil habitat—many practices help both at the same time—it should make many issues clearer. Not all of the aboveground discussion refers directly to management of soil, but most does. In addition, the practices we'll discuss contribute to one or more of the overall strategies: (a) growing healthy plants with strong defense capabilities, (b) stressing pests, and (c) enhancing beneficial organisms.

Aboveground Habitat Management

There are numerous ways that the aboveground habitat can be improved to help grow healthy plants, stress pests, and enhance beneficial organisms:

- Select crops and varieties that is resistant to local pests (in addition to other qualities such as yield, taste, etc.).

- Use appropriate planting densities (and companion crops) to help crops grow vigorously, smother weeds, and (with companion crops) provide some protection against pests. In some cases, blends of two or more varieties of the same crop (one susceptible to a pest but with a higher yield potential, and one that's resistant) have shown potential for increasing total yields for wheat and rice. Even though the farmer is growing the same crop, increased genetic diversity due to using different varieties (cultivars) seems to provide some protection. Perhaps there are possibilities for growing mixes of other crops as well.

- Plant perimeter (trap) crops that are more attractive to a particular pest than the economic crop(s) growing in the middle of the field and so can intercept incoming insects. (This has been successfully practiced by planting blue Hubbard squash on the perimeter of summer squash fields to intercept the striped cucumber beetle.)

- Create field boundaries and zones within fields that are attractive to beneficial insects. This usually involves planting a mix of flowering plants around or as strips inside fields to provide shelter and food for beneficials.

- Use cover crops routinely for multiple benefits, such as providing habitat for beneficial insects, adding N and organic matter to the soil, reducing erosion and enhancing water infiltration into the soil, retaining nutrients in the soil, and much more. It is possible to supply all of the nitrogen to succeeding crops by growing a vigorous winter legume cover crop, such as crimson clover in the South and hairy vetch in the North.

- Use rotations that are complex, involve plants of different families, and, if at all possible, include sod crops such as grass/clover hay that remain without soil disturbance for a number of years.

- Reduce tillage. This is an important part of an ecological approach to agriculture. Tillage buries residues, leaving the soil bare and more susceptible to the erosive effects of rainfall, and at the same time breaks up natural soil aggregates that help infiltration, storage, and drainage of precipitation. (The use of practices that reduce erosion is critical to sustaining soil productivity.)

Enhancing Soil Habitat

The general practices for improving the soil as a place for crop roots and beneficial organisms to thrive are the same for all fields and farms and are the focus of our discussions in the next chapters. However, the real questions are which ones are best implemented, and how are they implemented on a specific farm? These questions can only be answered by knowing the specific situation as well as the resources available on the farm. However, many practices are outlined below that may make the soil a better environment for growing healthy plants, stressing pests, and enhancing beneficial organisms:

- Add organic materials—animal manures, composts, tree leaves, cover crops, rotation crops that leave large amounts of residue, etc.—on a regular basis.

- Use different types of organic materials because they have different positive effects on soil biological, chemical, and physical properties.

- Keep soil covered with living vegetation and/or crop residues by using cover crops, sod crops in rotation, and/or reduced tillage practices. This encourages water infiltration, reduces erosion, promotes organisms that feed on weed seeds, and increases mycorrhizal numbers on the roots of the following crops.

- Reduce soil compaction to a minimum by keeping off fields when they are too wet, redistributing loads, using traffic lanes, etc.

- Use practices to supply supplemental fertility sources, when needed that better match nutrient availability to crop uptake needs. This helps to reduce weed and insect damage as well as pollution of surface and groundwaters.

- For soils in arid and semiarid climates, reduce salt and sodium contents if they are high enough to interfere with plant growth.

- Evaluate soil health status so that you can see improvement and know what other soil improving practices might be appropriate.

- Use multiple practices that improve the soil habitat. Each one may have a positive effect, but there are synergies that come into play when a number of practices—such as reduced tillage and cover crops—are combined.

Agroecology to Reverse Soil Degradation

The dramatic increase in crop production of the last 50 years has reduced the number of chronically undernourished people. However, these massive production gains have come at high environmental costs, which have affected soil and ecosystem health.

Currently agricultural policy is increasingly expected to face the combined challenge of producing sufficient food for a growing population while guaranteeing environmental restoration. Therefore, policy-makers are more frequently asked how to address the urgent need for soil and environmental restoration when millions of people are still hungry.

Niger - Keita.

Foods Security and Soil Degradation

"The world produces more than enough food to feed every member of the human family, yet 1 in 9 people do not have enough to eat". This was the opening sentence by the UN Secretary General, Ban Ki-moon, for the launch of EXPO 2015 in Milan, Italy.

Despite hosting almost all food production, rural areas also hold the majority of the world's food insecure people. Soils that are well managed by family farmers help ensure the four dimensions of food security: availability, delivering nutrients for crop growth; access, by improving family farm income through more reliable harvests; stability, by conserving water to support nearly year-round cropping; and utilization, by harvesting healthy nutritious food from healthy soils.

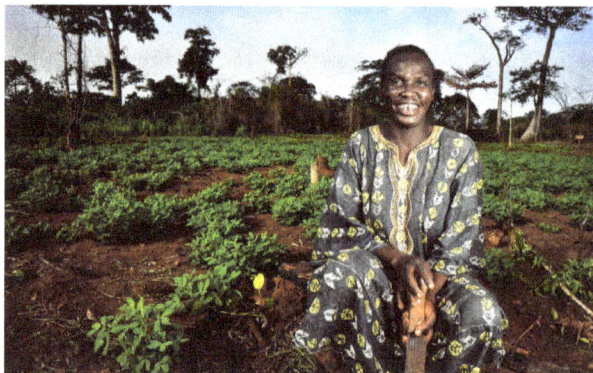

Central african republic - Dangala. Farmer in a peanut field.

Soil degradation consists of biological, chemical and physical degradation. Currently, about 33 percent of world soils are moderately to highly degrade. Forty percent of these soils are located in Africa and most of the remaining amounts are in areas that are afflicted by poverty and food insecurity. The strong relationship between soil health and food security calls for strategic and immediate actions especially at the local level to reverse soil degradation, in order to increase food production and alleviate food insecurity in the areas where it is most needed and in the context of climate change.

Agroecology as a Strategy to Reverse Soil Degradation

By understanding and working with interactions among soil, plants, animals, humans and the environment within agricultural systems, agroecology encompasses multiple dimensions of the food system, including ecological restoration, political and social stability and economic sustainability.

El salvador - A farmer holding two types of soil.

The agroecological approach starts by restoring soil life in order to re-establish and/or enhance the multiple soil-based biological processes. This requires:

- Increasing and monitoring soil organic matter: Soil organic matter is considered the most common deficiency in degraded soils and the main indicator for soil quality. Practical, accessible indicators can support local decisions and larger landscape monitoring and analyses for district level implementation.

- Facilitating and monitoring of soil biodiversity: Soil biological communities are directly responsible for multiple ecosystem functions.

- Build on local farmers' knowledge: Participatory scientific approaches to soil ecosystem management, such as Farmer Field Schools, are of great importance to inform farmers' knowledge with researchers' scientific principles in order better locally adapt agroecological systems.

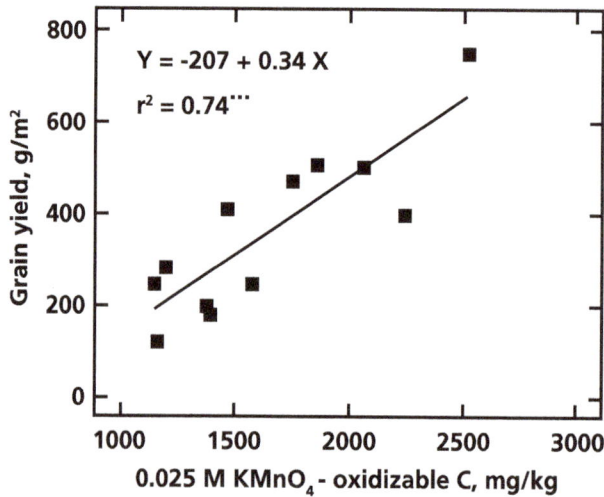

Relationship between corn grain yield and active fraction of soil organic matter.

LAOS - Nakhong Village. Integrated pest management training at a farmers' field school: collecting pest specimens in the field.

Agroecology as a Strategy to Restore Soils and Ecosystem Stability

Agroecology applies specific strategies based on temporal and spatial diversity, which guarantee local, stable and diverse year-round production and income. These strategies include:

1. Polycultures and agroforestry systems: The design of appropriate crop mixtures is more stable than monocultures as polycultures build on diverse crop resistance to soil pests and diseases and complementary uptake of soil nutrients and water in order to facilitate recycling of biomass and nutrients. The complementary traits of trees and crops enhance the efficiency of the whole systems, while litter mulch and the position of the trees along contour lines reduce erosion and soil degradation potential.

Sloping Agricultural Land Technology (SALT) is a specific agroforestry strategy in which annual and perennial crops are grown between contoured rows of leguminous species. SALT has been extensively tested and implemented in farmers' fields and experimental plots in Southeast Asia and has proven effective for reversing soil degradation while improving crop yields and farm's profitability.

Ghana - Wendu. Hedgerow intercropping. Hedgerow of Laucaena leucocephala (Leguminosae) and maize as a companion crop in a field at Wenchi. Leucaena fixes nitrogen.

2. Cover crops: Cover crops are usually leguminous crops grown to improve soil health by guaranteeing permanent soil cover, adding organic matter to soil and fixing atmospheric nitrogen. These help reverse soil degradation even in densely populated areas where long term fallows are simply no longer possible.

The use of Mucuna spp. as a cover crop in different African locations has increased soil organic matter, improved nitrogen availability in soils and positively affected yields.

3. Crop-livestock integration: Integrating livestock with crop production can tighten up nutrient cycles and diversify production, especially for smallholder family farms. In mixed farming systems, crop by-products are fed to livestock while manure is applied to cropland to sustain benefits from soil organic matter and nutrients availability.

In Ethiopia and Tanzania the design of mixed farming systems, which include multipurpose legume species such as Cajanus cajan (pigeon pea)–a drought tolerant multi-purpose legume–or Faidherbia albida –an indigenous leguminous nitrogen–fixing species with pods palatable for livestock, and leaves used as fertilizers-are well known to be effective in reversing soil degradation by controlling erosion, providing nitrogen-rich residues and increasing soil organic matter.

Time for Action

The design of diverse agroecological systems rooted in local ecological knowledge and based on system diversity and ecological synergies can significantly improve soil quality and reverse soil degradation while increasing the production of nutritious food.

Agroecology has already proven to be an effective strategy to address the global challenge that agriculture is facing as it accommodates the socio-political characteristics of food security with the need for restoring ecosystem functions.

El Salvador - Jocoaitique, Ladera area. Crop diversification: farmer working in maize plantation where henequen is also grown.

Agroecology is part of the Strategic Framework of FAO, in particular the Strategic Objectives of making agriculture, forestry and fisheries more productive and sustainable, increasing the resilience of livelihoods and reducing rural poverty. To facilitate a dialogue about Agroecology, its benefits, challenges and opportunities focusing at regional and national level, FAO is involved in regional conferences (held in 2015 in Latin America and the Caribbean, sub-Saharan Africa and Asia and the Pacific). Furthermore, FAO supports farmers' research networks to integrate scientific innovations with traditional farmers' knowledge.

Chapter 3

Agricultural Ecosystem

The functionally and spatially coherent unit of agricultural activity is known as an agricultural ecosystem. The biodiversity within an agricultural ecosystem is known as agrobiodiversity. All the diverse aspects of agricultural ecosystems as well as their management have been carefully analyzed in this chapter.

An agricultural ecosystem is an ecosystem managed with a purpose, usually to produce crops or animal products. Agricultural ecosystems are designed by humans, and are based on a long chain of experience and experiments. The emphasis in, for example, Western Europe has changed from maximum productivity only to also include environmental considerations, such as reduction of nutrient losses to groundwater and maintaining an open landscape with high biodiversity, etc. In less-productive regions such as sub-Saharan Africa, environmental considerations still have low priority.

Agricultural ecosystems comprise almost 40% (5 Gha) of the total land area of the Earth. About 11% of the total land area is arable land (cultivated with crops), and approximately 27% of the total land area is under permanent pasture, grazed by cattle, goats, sheep, camels, etc.

Plant biodiversity is extremely low – if weed control is successful there may be only one species present. In spite of this, belowground biodiversity can be high, although often lower than in natural ecosystems.

Ecological research performed in agricultural systems has advantages compared with research in most natural ecosystems, since agricultural fields are 'homogenized', that is, trees, larger stones, etc. are removed and regular soil cultivation to some extent evens out differences in topsoil properties over time.

Agroecosystems are increasingly expected to provide a gamut of ecosystem services including food production, C sequestration, nutrient recycling, and climate regulation. Simultaneously, agricultural intensification through the use of tillage, fertilizers, and pesticides alters soil physical, chemical, and biological properties, and thus increases yields at the cost of other ecosystem services. Tillage typically reduces mean SOC content, but also tends to homogenize the horizontal and vertical distribution of SOC. Since most soil biological communities are dependent upon SOC substrates, the spatial distribution of biologically mediated soil ecosystem services is impacted by agricultural practices that alter SOC. Fertilizer application has dramatic and predictable effects on the soil microbial community, generally leading to increased nutrient losses. Ecological intensification seeks to improve ecosystem service provisioning within agroecosystems by adding or modifying biological components of the system. Examples of ecological intensification include the use of cover crops, perennial plant species, and biochar soil additions. In the end, agricultural intensification must be combined with ecological knowledge and practices in order to balance crop yields with other critical ecosystem services.

Agroecosystem ecology is a multidisciplinary science that involves microbial, plant and animal ecologists, as well as those that work in above- and below-ground systems in both agricultural and natural/semi-natural contexts. All these protagonists use discipline-specific protocols, so networks tend to be carved up and dealt with piecemeal rather than as a whole, and how the components are interlinked over time and space is still poorly understood. Understanding the structure and dynamics of ecological networks that incorporate a wide range of interaction types is a growing area in ecology, partly driven by advances in computer modelling and novel molecular approaches, but ultimately by the desire to understand the real threat of biodiversity loss to ecosystem services and functioning. Much of the recent work in this area has been in agroecosystems.

For example, Pocock et al. linked plants with 11 groups of animals on an intensively studied organic farm in England, focussing on animals feeding on plants (butterflies and other flower-visitors, aphids, seed-feeding insects and granivorous birds and mammals;) and their parasitic dependants (primary and secondary aphid parasitoids, leaf-miner parasitoids, endoparasitoids of seed-feeding insects and ectoparasitoids of rodents). This study inevitably encompassed a wide taxonomic and functional range, which included animals regarded as bioindicators and as ecosystem service providers. Although this was just a subset of species interactions at the farm scale, the study was unique in attempting to connect multiple species-interaction networks, which have traditionally been studied in isolation. By examining topological 'robustness' as a measure of the tolerance of the network to species extinctions, Pocock et al. found that some networks (e.g. the plant–pollinator network) were far more fragile than others (e.g. the bird-seed-feeder network). They also found that robustness did not co-vary among them, suggesting that targeted management of one group will not necessarily benefit others. The study also used robustness values to compare the relative importance of plant species to the network's overall stability, and found that species such as thistles (Cirsium spp.), buttercups (Ranunculus spp.) and clover (Trifolium spp.) occurred in most habitats and were highly connected to many other species on the farm.

Evans et al. subsequently extended this approach to examine the effects of simulated habitat loss on the same spatial network of ecological networks. Habitat destruction is a primary cause of biodiversity loss and the impacts of management, such as habitat addition, loss and change, are likely to have large effects within ecological networks, as they will simultaneously affect multiple species across trophic levels. Evans et al. used

12 habitats (six managed and six non-managed) on the same organic farm to create multiple species-interaction networks for each habitat and then simulated sequential habitat loss under three scenarios: (a) random, (b) based on human decisions and (c) with a genetic algorithm to identify best- and worst-case permutations. Overall, the plant and animal groups exhibited high robustness, largely because habitats tended to have similar species composition and few unique interactions, despite considerable variation in management intensity and disturbance between habitats. Additionally, many of the animal groups (e.g. flower-visitors, birds and mammals) operated at spatial scales that integrated several habitats. These results suggest that the loss of a particular habitat may have little impact on animals, so long as suitable resources are available elsewhere. The models assumed that with the loss of a habitat-specific food source or host, animals could switch to alternate food sources in different habitats, but for some species this may not always be feasible. The models also assumed that the entire possible host range was observed: an observation likely to be affected by sampling bias. Despite these potential caveats, such novel analyses that incorporate environmental variation into the network clearly have considerable potential for predictive agroecosystem management and restoration.

Agrobiodiversity

Agrobiodiversity is the result of natural selection processes and the careful selection and inventive developments of farmers, herders and fishers over millennia. Agrobiodiversity is a vital sub-set of biodiversity. Many people's food and livelihood security depend on the sustained management of various biological resources that are important for food and agriculture. Agricultural biodiversity, also known as agrobiodiversity or the genetic resources for food and agriculture, includes:

- Harvested crop varieties, livestock breeds, fish species and non-domesticated (wild) resources within field, forest, rangeland including tree products, wild animals hunted for food and in aquatic ecosystems (e.g. wild fish);

- Non-harvested species in production ecosystems that support food provision, including soil micro-biota, pollinators and other insects such as bees, butterflies, earthworms, greenflies; and

- Non-harvested species in the wider environment that support food production ecosystems (agricultural, pastoral, forest and aquatic ecosystems).

Agrobiodiversity is the result of the interaction between the environment, genetic resources and management systems and practices used by culturally diverse peoples, and therefore land and water resources are used for production in different ways. Thus, agrobiodiversity encompasses the variety and variability of animals, plants and micro-organisms that are necessary for sustaining key functions of the agro-ecosystem, including its structure and processes for, and in support of, food production and food security. Local knowledge and culture can therefore be considered as integral parts of agrobiodiversity, because it is the human activity of agriculture that shapes and conserves this biodiversity.

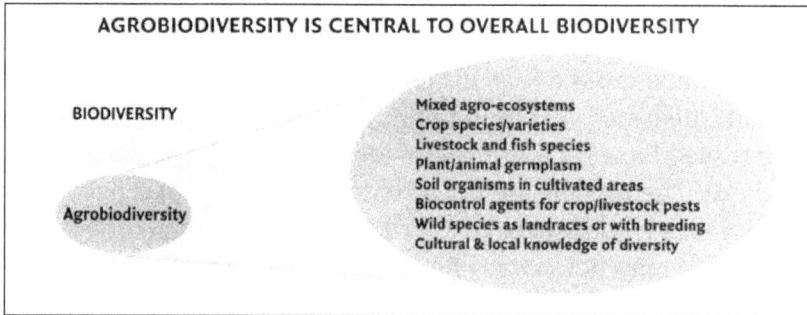

AGROBIODIVERSITY IS CENTRAL TO OVERALL BIODIVERSITY

BIODIVERSITY

Agrobiodiversity

Mixed agro-ecosystems
Crop species/varieties
Livestock and fish species
Plant/animal germplasm
Soil organisms in cultivated areas
Biocontrol agents for crop/livestock pests
Wild species as landraces or with breeding
Cultural & local knowledge of diversity

Agrobiodiversity is central to overall biodiversity

The variety and variability of animals, plants and micro-organisms that are used directly or indirectly for food and agriculture, including crops, livestock, forestry and fisheries. It comprises the diversity of genetic resources (varieties, breeds) and species used for food, fodder, fibre, fuel and pharmaceuticals. It also includes the diversity of non-harvested species that support production (soil micro-organisms, predators, pollinators), and those in the wider environment that support agro-ecosystems (agricultural, pastoral, forest and aquatic) as well as the diversity of the agro-ecosystems.

Many farmers, especially those in environments where high-yield crop and livestock varieties do not prosper, rely on a wide range of crop and livestock types. This helps them maintain their livelihood in the face of pathogen infestation, uncertain rainfall and fluctuation in the price of cash crops, socio-political disruption and the unpredictable availability of agro-chemicals. So-called minor or underutilized crops, more accurately, companion crops, are frequently found next to the main staple or cash crops. They often grow side by side and their importance is often misjudged. In many cases, from a livelihoods perspective, they are not minor or underutilized as they can play a disproportionately important role in food production systems at the local level. Plants that will grow in infertile or eroded soils, and livestock that will eat degraded vegetation, are often crucial to household nutritional strategies. In addition, rural communities, and the urban markets with which they trade, make great use of these companion crop species.

Collection of Wild Plants for Household Consumption

In Burkina Faso, and throughout the West African Sahel, rural women carefully collect the fruit, leaves and roots of native plants such as the baobab tree (Adansonia digitata), red sorrel leaves (Hibiscus saddarifa), kapok leaves (Ceiba pentandra) and tigernut tubers (Cyperus esculentus L.) for use in the families' diet. These supplement the agricultural grains (millet, sorghum) that provide only one part of the nutritional spectrum and may fail in any given year. More than 800 species of edible wild plants have been catalogued across the Sahel.

There are several distinctive features of agrobiodiversity, compared to other components of biodiversity:

- Agrobiodiversity is actively managed by male and female farmers.

- Many components of agrobiodiversity would not survive without this human interference; local knowledge and culture are integral parts of agrobiodiversity management.

- Many economically important agricultural systems are based on 'alien' crop or livestock species introduced from elsewhere (for example, horticultural production systems or Friesian cows in Africa). This creates a high degree of interdependence between countries for the genetic resources on which our food systems are based.

- As regards crop diversity, diversity within species is at least as important as diversity between species.

- Because of the degree of human management, conservation of agrobiodiversity in production systems is inherently linked to sustainable use - preservation through establishing protected areas is less relevant.

- In industrial-type agricultural systems, much crop diversity is now held ex situ in gene banks or breeders' materials rather than on-farm.

The Role of Agrobiodiversity

Experience and research have shown that agrobiodiversity can:

- Increase productivity, food security, and economic returns.
- Reduce the pressure of agriculture on fragile areas, forests and endangered species.
- Make farming systems more stable, robust, and sustainable.
- Contribute to sound pest and disease management.
- Conserve soil and increase natural soil fertility and health.
- Contribute to sustainable intensification.
- Diversify products and income opportunities.
- Reduce or spread risks to individuals and nations.
- Help maximize effective use of resources and the environment.
- Reduce dependency on external inputs.

- Improve human nutrition and provide sources of medicines and vitamins.

- Conserve ecosystem structure and stability of species diversity.

Spatial-temporal Mapping of Agro-ecosystems

Many scientists have conceptual problems differentiating (agricultural) land use information from information on other thematic ecosystem components, e.g. many maps, and other survey products, tend to confuse information on land use with, for instance, land cover. The impact of land use on land cover can hardly be studied if such products are used. It is a precondition for any exercise involving detection of change that causes (e.g. change in land use) and effects (e.g. change in cover) are kept apart. This principle is not open to compromise.

This paper discusses aspects to describe (agricultural) land use at plot level to optimize options to cluster, generalize and classify collected primary date, and to extrapolate the results spatially through modern RS/GIS techniques into map units that have an attached legend, in which the generalization or classification results are applied by theme.

Archiving properly collected primary land use data provides options to re-use them when new (e.g. monitoring) studies are called for. Different, study-specific classification rules can then be applied on the available primary data for alternative clustering, generalization, classification and extrapolation.

The approach followed here is intended to be both practical and conceptually correct. The bottom-up approach that was adopted leads to the holding-level where actual decision-making by individual land users takes place. Studies of biophysical land use system performance generates inputs for socio-economic evaluation, culminating in the definition of planning scenarios that conserve land resources and are rewarding for both primary and secondary stakeholders.

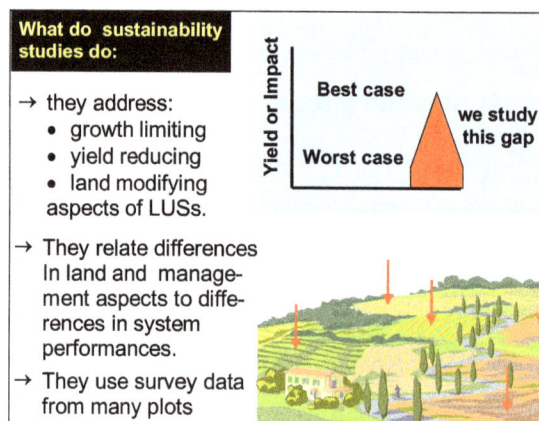

Sustainability Studies

Agro-ecosystem (land use system) studies must include studies of the land. Management activities (operations) at plot level aim at modifying one or more aspects of land, e.g. the soil, flora/fauna, or infrastructure. Operations are carried out to support one or more land use purpose(s), e.g. to harvest a good crop but they can also have negative side effects that affects the sustainability of the system. Often, operations are pre-planned, but they can be of a remedial nature depending on dynamic land processes, for example, incidence of pests and diseases, weeds infestation, water and nutrient deficiencies, etc.

Programmes or projects that address the stated sustainability issues specifically require timely and reliable (spatial) information on the productivity and sustainability of current agricultural land use systems. However, there is a general paucity of land use information in many developing countries and it is often difficult for the range of potential clients to access the information that is available. Young refers to the described vacuum as:

- "To an extent which, viewed in retrospect, is remarkable, methods for the collection and analysis of land use data have lagged behind those for natural resource surveys", and
- "The situation with respect to land use classification was comparable with that for soils in about 1950: a large number of systems devised for national use, with no guidelines for comparison".

Whilst:
- "At national level, many countries are now seeking to monitor land use change as a basis for policy guidelines and action", and "land use is generally treated as the second most fundamental set of statistics, following population".

In short, we need good land use data to address questions as put on record by the UNCED conference in Rio, e.g.:
- To identify options to solve future food requirements.
- To understand and combat environmental degradation. and, we need practical concepts and approaches to:
  - Gather, manage, classify and map land use information.
  - Study various aspects of present day land use systems.

Information Technology for Sustainable Land Management

Geomatics

In the early nineties, a 'think-tank' of the Atlantic Institute, representing faculties from NE-USA and E-Canada, came to the conclusion that:

- Trends in land management studies are towards geomatics, defined by the Atlantic Institute as "the scientific management of spatial information". Boundaries between formerly separate disciplines have become increasingly fuzzy;

- Developments have moved from a period of innovation (1960-1980: technology driven, little data) through a period of integration to a period of proliferation (1990 -: systems integration, mass dissemination, information customer driven).

Information technology (IT) facilitates integration of information processing. This is obvious from the advent of management decision support systems that grew out of the management information systems. Geographic Information Systems (GIS) are a direct result of this integration. GIS provide the user community with tools that are unprecedented in their potential and challenge existing facilities. IT also has the capability to transform a data set at relatively low cost into new information products for specific users. An important consequence of these integration and customization characteristics of IT is that combined processing of data sets can deliver new information products with an added value over the source data sets. IT has particular significance for interdisciplinary land use planning. It facilitated decentralization of governance and progress in communication, it spurred research into sustainable use of natural resources, and it opened international markets for technology and knowledge.

Quality of Present Day Land use Information Systems

Stakeholders report that the effective use of GIS technology is constrained by the limited adequacy of data on land use systems. The constraints were recorded at selected (sub-) national institutes in a number of developing countries and in four European countries.

The recorded statements on present day land use system information for natural resource management and planning called for (guidelines on) data harmonization. Aspects to be considered are listed in table.

Table: Constraints regarding effective use of land use system information as reported by stakeholders.

Data Aspect	Problem	Frequency
Availability	What is where?	Occurs
(supply defined?)	Unobtainable, restricted	Often
	Limited coverage	Regular
Format	Supplier defined	Often
	Data integration problems	Often
	Different parcel registries	Regular
Quality	Lack of uniformity	Often
	No accuracy assessments	Regular
Documentation	Often incomplete	Occurs
	Poor nomenclature	Often
Geo-referencing	Sometimes absent	Occurs

Costs	Often expensive	Regular
Updates	Poor update frequency	Regular
	No time series	Occurs
Coordination	End users not involved	Often
	Poor between organizations	Often
	No regulations	Occurs
Classification	Not tailored to user needs	Often
	No user consultations	Often
	Lack of uniformity	Constant
	Limited utility	Constant

Table: Data aspects that need attention if the quality of present day land use system information is to be improved.

Data Aspect	Problem
Concepts	Differentiate between land use and land cover
Data accuracy and consistency	Survey methodologies
Scale and legend correctness	Observation units
Type of data	Classes vs. Numeric information
Class definitions	User consultations
Definitions	Nomenclature
Consistency for time-series	Replicability
Data formats	Relational database, GIS formats
Documentation	Set regulations

Land use Information System Concepts

At present, digital databases are being developed that form a part of the information infrastructure required for sustainable land management at various scales. The development of GIS has dramatically increased the demand for reliable geo- referenced data at all levels of detail.

UN organizations supported by specialized institutes develop standards and software for the collection and analysis of geo- referenced information on climate, soil and terrain conditions, water resources, land use, land cover and bio-diversity, and on social and economic conditions. All of these must be referenced with up-to-date and accurate topographic and cadastral information. (Inter-) national programs are needed to unite such databases in a uniform geo-information infrastructure. Examples of initiatives are GSDI (Global Spatial Data Infrastructure), EUROGI and Eurostat (Europe), and RAVI (The Nether- lands). The Open GIS Consortium is a similar initiative on a world scale that was started by US-based GI-industries.

Figure provides a basic "root structure" (concept) of a comprehensive land use systems

(LUS) information system. Given that it was developed from the conceptual LUS-diagram provided in figure, the part on land use is elaborated in further detail.

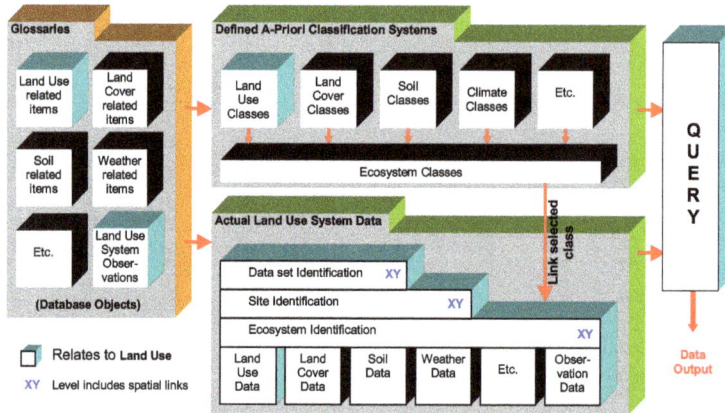

The basic "root concept" (structure) of a LUS- information system.

Figure shows that a land use system is composed of two main elements: land and land use. A Land Use System (LUS) was defined by De Bie as: "A specific land use, practiced during a known period of time on a known unit of land that is considered homogeneous in land resources"; Land Use was defined as: "A series of operations on land, carried out by humans, with the intention to obtain products and/or benefits through using land resources". Land use purpose(s), i.e. the intended products or benefits of land use, and an operation sequence, i.e. a series of operations on land in order to realize one or more set land use purposes, characterize land use. Note that land use systems have both spatial and temporal dimensions. These must be understood if one endeavors to describe, classify, survey or study land use systems at the level of spatial aggregation required for solving specific natural resources management problems.

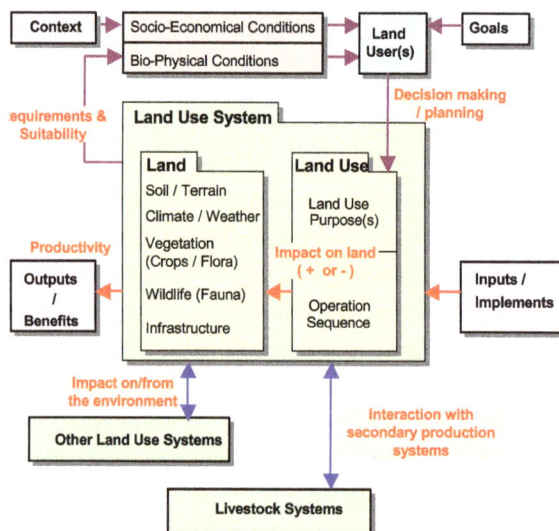

Conceptual Structure of a Land Use System (LUS) providing Practical 'Study Entries'.

For the land use part, figure suggests two sets of relational database files as required to capture land use descriptions. The first set is called "Actual Land Use System Data" and contains collected land use information, either primary or secondary. The second is called "Defined A-priori Classification Systems" and contains information on a-priori land use classes, i.e. class names and classifiers used to define the classes. Parameter values used in each set are derived form a "Glossary". The data- base files are all linked through index-keys; relevant links are presented in figure. The various data files are:

- Data set identifications: Contains general information that identifies a particular data set, including the administrative area; project under which the survey takes place, names of enumerator and respondent, holder, etc.

- Site Identifications: Contains data that provide detailed information about the geographic location of the site(s) under study such as map unit, cadastral no., parcel size etc.

- Land Use System Descriptions: Contains general information about the land use system such as plot location and size, operations seq. duration, a-priori land use class, etc.

- Operations and Observations: Contains data on individual operations and observations.

- Land Use Classes: Contains information on a-priori land use classes. A land use class is defined without any temporal and spatial dimensions. It is a universally applicable land use description based on well-defined classifiers.

To understand the "operation sequence" better, some definitions follow:

Operations are intended to modify land aspects, e.g. soil characteristics or land cover. Some modifications are permanent (constructing infrastructure) whereas others can be of a temporary nature, e.g. the successive land cover types 'bare soil, crop, and stubble' are brought about by 'ploughing, planting and harvesting'. Impacts of operations may exceed the in- tended effects resulting in, e.g. erosion, accumulation of pesticide residues, loss of soil fertility, etc. Four basic types of impact can be distinguished; they relate to soil/terrain, flora/fauna, infrastructure and air.

Observations are defined as: "A record of one or more land conditions that are relevant to the performance of a land use system." Examples of observations are "water shortage during crop establishment", or "recorded limitation of the rooting depth of crops". Observations can be made at any moment during the life span of the land use system; the land user makes them often and information about such observations is obtained through interviews. Observations frequently provide important information on the temporal properties of the land use system; such information is not stored in databases that contain only static or generalized data on land.

Use of RS for Land use Mapping

The spatial characteristics of a land use system define its boundary. For agricultural purposes, a land use system can be limited to a plot. A plot was defined as "A piece of land, considered homogeneous in terms of land resources and assigned to one specific land use."

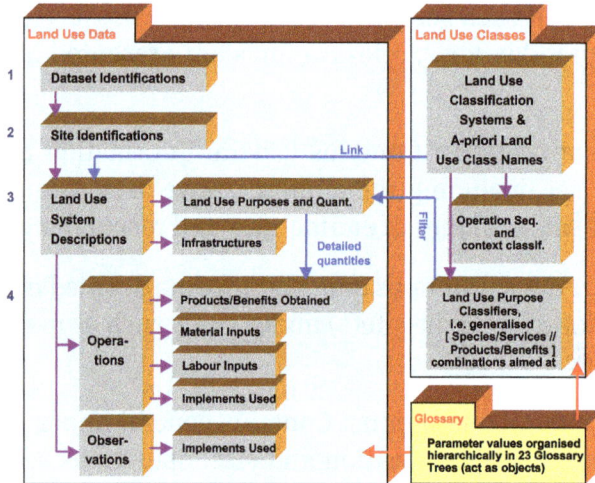

The underlying database files on Land Use (4 main levels); Squares represent one database file each, links from the Glossary files to the ; Land Use Data and Land Use Classes data files are not shown in detail.

The "operation sequence" is an essential component of any crop calendar. A crop calendar was defined as: "A sequential summary of the dates/periods of essential operations, including land preparation, planting, and harvesting, for a specific land use; it may apply to a specific plot, but is frequently generalized to characterize a specified area." Plot specific crop calendars form the key to map land use with the support of (multi-temporal) RS-imagery.

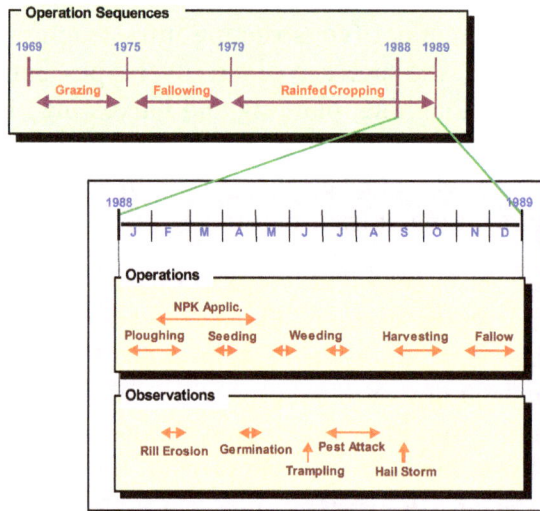

Illustrating Land Use Operations and Observations.

A cropping pattern is traditionally defined as: "The yearly sequence and spatial arrangement of crops or of crops and fallow on a given area". In view of the crop calendar definition, the cropping pattern definition can be sharpened to: "The spatial and temporal arrangement of crops (trees) on a specific plot." Generally, a cropping pattern refers to a period of one year, but may also contain information on crop rotation. The definition contains spatial information (within a plot) that is not present in a crop calendar, but lacks actual date/period references as provided by a crop calendar. Cropping pattern terminology is area a-specific and therefore often used to classify land use. Legends of land use maps will considerably improve when cropping pattern syntax is used.

Land use is dynamic and operations and events take place that the surveyor cannot personally witness. The land user has that capability and is (assumed) able to recall important land use system aspects for at least the past growing season. His/her knowledge regarding those aspects can only be sampled through interviews. For land use surveys, interviews at plot level are thus essential. To complement and verify interview information, observations by the surveyor must be recorded on a field form. Interviews can be based on questionnaires or on checklists; each has its respective strengths and weaknesses. Use of checklists is recommended for scientific research to prevent that relevant site-specific operation data or occurrences are overlooked.

Table: Overview of 'Information Data Sets' with Specific Relevance for Land Use Surveys.

RS-Image characteristics

•   1D-features (tone, color), as related to:
     ○ Crop calendars, cropping patterns and other land use operations.
     ○ Infrastructure.
•   2D-features, such as:
     ○ Field sizes, shapes and patterns.
     ○ Internal patterns (textures, grids, mottles).
     ○ Line features.
•   3D-features (on APs): vertical structure, no. of layers.

Observation/measurement data:
•   Plot size, coordinates, slope, position, etc. •   Crops (residues) and infrastructure present in / around the plot. •   Land cover data (crop condition, growing stage, weed incidence, biomass, height, etc.) •   Ground cover status (bare soil, mulch, crop residues, etc.) •   Specific observations (soil characteristics, tillage condition, erosion status, hydrological aspects, pests / diseases incidence, evidence of grazing, etc.)
Interview data:

- Holding/holder information (profile).

- Site aspects (tenancy arrangement, cadastral no., distance to holding).

- Land use system (plot) aspects for the period considered:

- A-priori land use class.

- Crops grown / services provided (% of area, numbers, etc.)

- Land use purposes.

- Operation aspects (the crop calendar and cropping pat- tern):

  ○ Operation name; species involved; % of plot in- volved; period / periodicity / dura- tion and task times; main power source.

  ○ Labor and material inputs and implements used.

  ○ Products / benefits obtained.

- Observations by land user and indigenous knowledge:

  ○ Soil related (workability, infiltration rate, fertility status, etc.)

  ○ Weather related (hail storm, dry period, etc.)

  ○ Crop related (pests, diseases, lodging, wilting, etc.)

Land use Classification Concepts

Classifiers

There is enormous variation in land use worldwide. To map land use, compile land use statistics, and carry out land use planning, common characteristics in the wide variety of land uses must be identified. Common land use characteristics can be identified in two ways:

- By generalizing the descriptions of actual land use systems to a description that conforms to, e.g. land use names/descriptions in map legends; these descriptions hold only for specific areas and periods of time.

- By classification of land use descriptions resulting in descriptions that are not limited to a certain area or time frame.

Land use classification was defined as: "The process of defining land use classes on the basis of selected diagnostic criteria", and a land use class as: "A generalized land use description, defined by diagnostic criteria that pertain to land use purpose(s) and operation sequence followed; it has no location or time indications." Land use classes are exclusively based on attributes of land use in the context of a LUS.

Classification (of land use) must be based on unambiguous diagnostic criteria that are known as "classifiers". Often classifiers are not properly documented in land use (classification) reports; only names of classes are given.

A land use class is a taxon that is solely based on information on land use purpose and operation sequence. In combination with attributes of land, the land use class becomes extended to a LUS-class. Using LUS-classes does not allow assessing the suitability of a certain land unit for a certain land use or for monitoring land use changes. In spite of this, land characteristics are sometimes considered as classifiers, resulting in land use system classes such as "un-used bare soil" or "protected tropical forest".

Three types of classifiers can be applied to define land use classes:

- Land use purpose classifiers: specify aimed at [Species/Service - Product/Benefit] combinations in general terms. At least one combination must be specified for each land use class. No new products or benefits can be added to define sub-classes, but existing definitions can be sharpened or split into several new definitions.

- Land use operation sequence classifiers: specify (one or more) aspects of operations in general terms. For sub-classes new classifiers can be added; higher level classifiers remain valid for all sub-classes, or can be further narrowed down.

- Land use context classifiers: specify (one or more) circumstantial aspects of the land use in general terms that are not a part of the land use purpose or operation sequence. Context classifiers are better not used but are mentioned to link up with existing practices. They can include: Land aspects, e.g. infrastructure, tenancy arrangements, etc.; Holding (context) aspects, e.g. origins of inputs/ implements, destinations of outputs (market orientation), capital intensity, holder attitude, goals of holder, credit availability, pricing policies, etc.

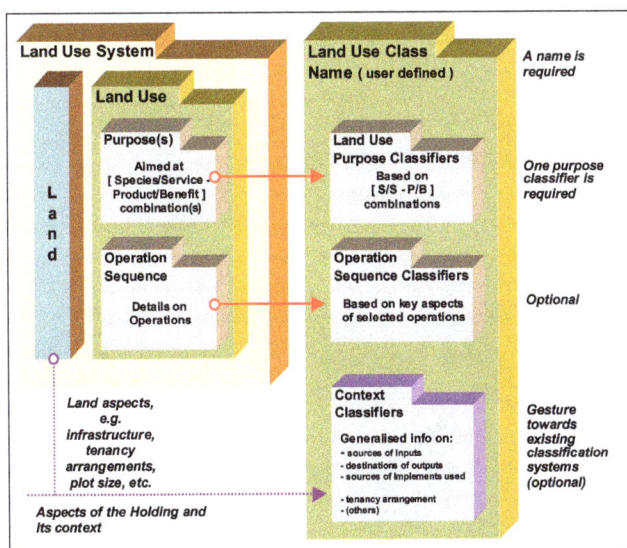

Classifiers used to define a land use class.

The parametric method of defining land use classes employs a combination of classifiers to define a land use class. Table presents an example of a land use class defined in terms of independent classifiers. The provided list is not exhaustive but is intended to "grow" into a standard set of classifiers for use in the preparation of land use classification systems. In addition, use of classifiers is helpful for merging of classification systems, and to correlate classes defined under different classification systems.

Table: Example of a land use class definition.

Codes	Shifting Cultivation
n.a. n.a.	**Purpose Classifiers** • Plants for plant produce, and • Animals for animal produce.
A.1.1.2.1.4 B.1.4 F.0 I.2 K.1 L.1	**Operation Sequence Classifiers** • Agricultural production • Crop production • Temporary (arable) cropping Multiple cropping Intercropping
	Patch ~, and • Extraction / Collection • Yes • Mix of hunting and vegetation exploitation, and • Recreation and tourism • none, and • Cultivation factor (R) • R < 33%, and • Main power source for tillage • manual power only, and • Material inputs • low.
cA.0 cB.0 cF.0 cG.1 cI.0	**Context Classifiers** • Tenancy arrangements / Land rights • Taken in possession, and without a secure title, and • Connectivity • poor, and • Market orientation • subsistence, and • Capital intensity • low, and • Secondary Infrastructure requirements • none.

A land use classification system was defined as: "A structured set of land use class definitions." Most land use classification systems are hierarchically structured and obey the following rules:

- The defined land use classes are mutually exclusive at each level, and
- Classes at sub-levels are a further specification of a class at a higher level.

A-priori versus a-posteriori land use classification:

- A-priori classification: Implies that land use classes are defined before collecting the actual data. Classifiers used are based on expert knowledge, study objectives, or conform to classes defined by international organizations, national institutions etc. The main advantage of a-priori systems is that classes are standardized. Assigning class names to land use descriptions is called "identification".

- A-posteriori classification: Means that land use classes are defined using classifiers that are based on (analysis of) data collected. The advantage is that classifiers can be defined that fit recorded study results.

Harmonizing Classifiers

The growing demand for global assessment of land use (possibilities) generated a need for a universal classification system. Many attempts to develop a comprehensive classification sys- tem have been made. Fresco et al. concluded that: "Yet, there is no satisfactory and commonly accepted method of defining and classifying land use globally, let alone a definition of the major classes of land use as such. This situation thwarts the systematic collection of data pertinent to use classification".

Development of a comprehensive classification system for land use is still far away. Earlier efforts were all discontinued, and there is growing recognition that different land use studies re- quire different classification systems pending on set objectives, areas studied, and methods followed. For example: if remotely sensed images are used to map land use, classifiers used are strongly correlated with land cover whereas land use studies that center around farming system analysis will rather base their class definitions on land use purpose(s), labour inputs, etc. Each study can independently select the level at which a particular classifier is used, e.g. 'irrigated' can be a classifier at the highest level, or at any lower level, or can simply not be used.

If one universal classification system is a practical impossibility, then the problem remains that many classification systems remain in use with different classifiers at different levels. Standardization of land use classifiers would allow correlation of land use classes used in different studies. This standardization would keep the possibility to prepare user-defined classification systems open and not compromise the possibility to compare existing classification systems. It would then be possible to cross-tabulate different sets of land use classes to study their mutual (dis-) agreement.

The various criteria used around the globe to define classes form the basis to adopt an actual 'reference system'. They are the 'bridge' that can be used to compare and translate defined classes; it is thus essential that the criteria used are documented and existing classification systems are studied to define the 'basic set' of criteria.

Agro-ecosystem Analysis

The health of a plant is determined by its environment. This environment includes abiotic factors (i.e. sun, rain, wind and soil nutrients) and biotic factors (i.e. pests, diseases and weeds). All these factors can play a role in the balance, which exists between herbivore insects and their natural enemies. If we understand the whole system of interactions, we can use this knowledge to reduce the negative impact of pests and diseases.

Decision making in Integrated Pest Management requires a thorough analysis of the agro-ecosystem. Participants in IPM training will have to learn how to observe the crop, how to analyze the field situation and how to make the proper decisions for their crop management. This process is called the Agro-Eco-System Analysis (AESA).

AESA involves three steps:

- Observation
- Analysis
- Decision making

When participants of IPM training learn to do an agro-ecosystem analysis (AESA) they will make a drawing on a large piece of paper, in which they include all their observations. The advantage of using a drawing is that it forces the participants to observe closely and intensively. It is a focal point for the analysis and for the discussions that follow, and the drawing can be kept as a record.

AESA Methodology

The following methodology is taken from a guide on IPM training where the participants were learning to do an AESA in rice. For other crops, the approach could be slightly different, but the basics are the same.

Participants in a training making field observations

- Go to the field in groups. Walk across the field and choose 10 plants randomly. Observe keenly each of these plants and record your observations:

  ○ Plant: observe the plant height, number of tillers, crop stage, deficiency symptoms, etc.

  ○ Pests: observe and count pests at different places on the plant.

  ○ Defenders: observe and count parasites and predators.

  ○ Diseases: observe leaves and stems and identify any visible disease symptoms.

  ○ Rats: count numbers of plants affected by rats.

  ○ Weeds: observe weeds in the field and their intensity.

  ○ Water: observe the water situation of the field.

  ○ Weather: observe the weather condition.

- While walking in the field, manually collect insects in plastic bags. Use a sweep net to collect additional insects. Collect plant parts with disease symptoms.

- Find a shady place to sit as a group in a small circle for drawing and discussion.

- Kill the insects with some chloroform on a piece of cotton.

- Each group will first identify the pests, defenders and diseases collected.

- Each group will then analyze the field situation and present their analysis in a drawing (AESA drawing).

- Each drawing will show a plant/hill representing the field situation. The weather condition, water level, disease symptoms, etc. will be shown in the drawing. Pest insects will be drawn on the left. Defenders (beneficial insects) will be drawn on the right. Write the number next to each insect. Indicate the plant part where the pests and defenders were found. Try to show the interaction between pests and defenders.

- Each group will discuss the situation and make a recommendation.

- A member of each group will now present their analysis in front of all participants. Make sure that a different person will present each week.

- The facilitator will facilitate a discussion by asking guiding questions.

- The facilitator also makes sure that all participants (also shy persons or illiterate persons) become actively involved in this process.

- Formulate a common conclusion. The whole group should support the decision on what field management is required.

- Make sure that the required activities (based on the decision) will be carried out.

- Keep the drawing for comparison in the following weeks.

Agroecosystem Role in Climate Change Mitigation and Adaptation

Modern agriculture has been hugely successful in meeting global demand for food, fiber and feed, and has recently made impressive strides in the production of liquid transportation fuels. Agriculture productivity was achieved through a vast transformation of the Earth's surface on 15 million km2 of cropland and 28 million km2 of pasture. This change in land-use has resulted in impacts on a global-scale to ecosystem structure and function, including reductions in critical ecosystem services that moderate climate, air pollution, water contamination and soil degradation. New constraints may amplify these impacts as the agriculture system confronts rising production costs, accelerating climate change and increasing scarcity of natural resources.

An emerging response to the diverse impacts from the agriculture system has been to turn to ecologically sound agriculture practices that may enhance environ- mental quality, while also improving social and economic sustainability. Conservation tillage, cover cropping, intercropping and integrated pest management are among the many strategies employed. One underlying theme in developing the scientific basis and design of these agroecological practices is the biotic interactions that determine agroecosystem function. Managing biotic interactions in farm systems may reduce or eliminate the need for the very external inputs, which drive pollution, land degradation and the loss of biodiversity. Global adoption of such low-input practices has risen dramatically owing to new market opportunities for organic foods, as well as government incentives and mandates.

Agroecological approaches to farming represent a transformative change to many existing agriculture systems. Such a dramatic shift in agriculture practices may incur tradeoffs. One tradeoff highlighted in recent work is the potential for low-input agricultural systems to result in greater GHG emissions than the farming systems that they replace. The climate change dimension of agriculture sustainability is of great concern since on-farm GHG emissions are currently 5.1–6.1 billion tons CO_2-equivalents y^{-1}, which is 10–12% of total anthropogenic GHG emissions. In addition to on- farm emissions, agriculture is responsible for significant emissions in other categories reported in GHG inventories including deforestation, as well as industrial and energy sectors through the production of fertilizers, pesticides, herbicides, machinery and electricity. It is certain that low-input systems will reduce the emissions associated with fossil fuel intensive inputs. At the same time, there exists a potential for increased GHG emissions if the low-input systems are assumed to result in a decrease in crop yields, which drives an increase in agriculture areas and deforestation (extensification). Recent work on this tradeoff concludes that low-input agriculture systems result in greater net GHG emissions than high-input systems, since the increase in emissions from extensification is greater than the decrease in emissions from reduced inputs. The central assumption that a global-scale adoption of low-input systems would lead to extensification is equivocal. Nevertheless, the results of this study point to a critical knowledge gap where the role of agroecological systems in global GHG emissions as well as other potential synergies and tradeoffs with climate change mitigation and adaptation are uncertain.

The difference in yields between agroecological farming practices and the systems they replace has long been a subject of debate. The yield question is framed within the context of food security and the ability of agroecological farming systems to meet the growing global demand for agricultural goods. Some question whether this is relevant given other barriers to equal food access. Regardless of the relevance of yields to food security, yields appear to be important to the question of net GHG impacts of agriculture. In general, agroecological approaches may reduce yields in developed countries and increase yields in developing countries, but the impact on global production is unclear. Furthermore, it is uncertain how yields will compare for future climate regimes. There is evidence that agriculture practices designed with agroecological principals may be more robust to the changing climate system and thus yields may benefit from these practices . In this sense sustainable agriculture is an important climate adaptation measure. Another central uncertainty is the spatial distribution of the yield gaps. The spatial gradients in the yield gap will contribute to the spatial distribution of extensification, which in turn determines the magnitude of the GHG emissions associated with land use change. Given the importance of yield to determining the climate impacts of agriculture, there is a great need for spatially diverse farm experiments as well as geospatial analysis incorporating global land-use models and ecosystem carbon storage.

While the question of yields has been argued to be paramount for understanding GHG emissions from agroecoystems, emerging issues in global agroecology point to climate change mitigation from agroecosystems of equal or greater magnitude. In additional to GHG emissions, agriculture contributes to climate forcing through the exchange of water and energy between the land and atmosphere. This so-called bio-physical effect includes changes in the amount of sunlight absorption, water evapora-tion from plants and the soil, the roughness or unevenness of the vegetation canopy, and the production of convective clouds and rainfall. A recent study of the biophysical effects of perennial agriculture systems finds substan- tial climate cooling from in-creased evaporation that is significant at regional and global scales. While this study focuses on low-input biofuels cropping systems, the results suggest that investiga-tions of food systems have also overlooked the beneficial biophysical climate effects of agroecosystems. Furthermore, the biophysical cooling effects of agroecological farm- ing practices would likely have an indirect impact on yields. Increasing tem-peratures may lead to a nonlinear degradation in crop yields. Thus, the biophysical cooling associated with agroecological farming practices would tend to reduce tem-peratures and ultimately moderate degradation in yields. The biophysical effects of agroecosystems on climate and regional yields are largely unknown. The scientific basis for these effects could be developed through a combination of integrating eddy flux measurements into trials and integrating agroecological management into re-gional climate models.

Even if the net effect of GHG and biophysical forcing from agroecological farming prac-tices yields a warmer climate, the climate impact needs to be carefully evaluated within an integrated assessment of agroecosystems. A warming climate is not an impact in and of itself. The problems that stem from climate change are in many cases the very ecological, social and economic problems that agroecological practices address. For example, climate change will likely lead to more con- centrated precipitation events, which will degrade water quality while agroecological design can reduce non- point source pollution from runoff, sediment loss and leaching of nutrients and pesticides. Thus, simply knowing that an agriculture approach contributes to climate change is not sufficient to provide an assessment of sustainability.

There is a need for a more careful examination of GHG emissions from agroecosys-tems, for new investigations of biophysical climate impacts and for the inclusion of these studies into an integrated assessment of agriculture sustainability. The inte-grated assessment in particular is a formidable scientific challenge, which would bring together multiple disciplines as well as farmer / researcher collaborations to ad-dress key dimensions of agroecosystem sustainability. It is perhaps fitting that such a transformative approach to agriculture would require an equally transformative approach to science.

Development of Sustainable Agroecosystems

Since its beginning, agriculture has tested the resiliency of nature. Natural communities have been replaced with artificially supported, productive communities. Agriculture's objectives generally have been to achieve maximum yields, operate with a maximum profit, minimize year-to-year instability in production, and prevent long-term degradation of the productive capacity of the agricultural system. Theoretically, these objectives should be compatible and mutually reinforcing. Unfortunately, developments in agriculture have removed the crop ecosystems from their parent nonagricultural ecosystems s to the extent that agroecosystems and natural ecosystems have become strikingly different in structure and function.

The maintenance of an imposed order of simplified agricultural systems against the natural tendency toward entropy, diversity, and stability demands energy and resources. The depletion of nutrients, loss of soil fertility, and the alteration of soil structure must be compensated for by large subsidies of fertilizer and soil conditioners. Similarly, pesticides must be applied to compensate for the lack of self-regulating mechanisms in monocultures. The large-scale structural changes that have been made in agriculture include the creation of large, specialized farms, increased mechanization and use of biochemicals, regional specialization of production, and increased interregional marketing.

Modern farming has thus become a highly complex activity, where gains in crop yield depend directly on intensive inanagement and on the uninterrupted availability of supplemental energy and resources. This approach is no longer appropriate in an energy-troubled era; therefore, progress towards a more selfsustained, energy-efficient agriculture is desirable. However, when examining the problems that confront the development and adoption of sustainable agroecosystems, it is impossible to separate the biological problems of practicing "ecological" agriculture from those of inadequate credit, technology, education, political support, and access to public service. Social complications, rather than technical ones, are likely to be the major barriers against any transition from high capital/energy production systems, to labor-intensive, low energyconsuming agricultural systems.

Modern agriculturalists had assumed that the agroecosystem/natural ecosystem dichotomy need not lead to undesirable consequences, yet, unfortunately, a number of "ecological diseases" have been associated with the intensification of food production. They may be grouped into two categories: diseases of the ecotope, which include erosion, loss of soil fertility, depletion of nutrient reserves, salinization and alkalinization, loss of fertile croplands to urban development, and diseases of the biocoenosis, which include loss of crop, wild plant, and animal genetic resources, elimination of natural enemies, pest resurgence and genetic resistance to pesticides, chemical contamination, and destruction of natural control mechanisms. Under conditions of intensive management, treatment of such "diseases" requires an increase in the external costs to the extent that, in some agricultural systems, the amount of energy invested to produce a desired yield surpasses the energy harvested.

Restoring Ecological Health

To restore "ecological health" in agricultural systems energy and resource overuse should be curtailed, production methods that restore community stability should be employed, a maximum of organic matter and nutrients should be recycled, the best possible multiple use of the landscape should be made, and efficient energy flow should be ensured. Also, as much food as possible should be grown locally that is adapted to the local environment and local taste.

The technical development of such systems must contribute to rural development and social equality. For this to occur, political mechanisms must encourage substitution of labor for capital, reduce levels of mechanization and farm size, diversify farm production, and emphasize worker-controlled enterprises. Social reforms along these lines have the added benefits of increasing employment and reducing farmers' dependence on government and the pressures of credit demands and industry.

We recognize that these proposed changes conflict with the Western capitalist view of modern agricultural development. It may be argued, for example, that increased mechanization reduces production costs or is necessary in areas where adequate labor is unavailable and that diversified production creates problems for mechanization. Another concern is that sustainable technologies will fail to feed as many as 2 billion additional people by the close of this century. Each of these criticisms is valid if analyzed within the current socioeconomic frame work, but not so if we realize that the proposed sustainable agroecosystems represent profound changes that would have major social and political implications. We believe that most of the present and future problems of malnourishment and starvation are due more to patterns of food distribution and political economics than to agricultural limits or the type of technology used in food production.

Guidelines for Achieving Dynamic Stability

Productivity in agricultural systems cannot be increased indefinitely. A ceiling is placed on potential productivity by the physiological limits of crops-the "carrying capacity" of the habitat and the external costs incurred from the efforts to enhance production. This point is the "management equilibrium" where the ecosystem, considered being in dynamic equilibrium with environmental and management factors, produces a sustained yield. The characteristics of this balanced management will vary with different crops, geographical areas, and management objectives, and, therefore, they will be highly "sitespecific." However, general guidelines for designing balanced and well-adapted cropping systems may be gleaned from the study of structural and functional features of the natural or seminatural ecosystem remaining in the area where agriculture is being practiced. Four major sources of "natural" information can be explored.

Primary Production

Depending on climatic and edaphic factors, each area is characterized by a type of vegetation that has a particular biomass production capacity. An area covered by natural grassland (i.e., of a standing crop value of 6600 g/m2) is not able to support a forest (i.e., 26,000 g/m2) unless external subsidies are added to the system. It follows, then, that if natural grassland needs to be transformed into an agricultural system, it should be replaced by cereals rather than by orchards.

Table: Structural and functional differences between natural ecosystems and agroecosystems.

Characteristic	Agroecosystem	Natural ecosystem
Net productivity	High	Medium
Trophic chains	Simple, linear	Complex
Species diversity	Low	High
Genetic diversity	Low	High
Mineral cycles	Open	Closed
Stability (resilience)	Low	High
Entrop	High	Low
Human control	Definit	Not needed
Human control	Short	Long
Habitat heterogeneity	Simple	Complex
Phenology	Synchronized	Seasonal
Maturit	Immature, early successional	Mature, climax

Land use Capability

Soils have been classified into eight land use capability groups, each determined by physicochemical factorsslope, water availability, etc. According to this classification, soils of class I and II have a high natural fertility, good texture and permeability, and are deep, erosion-resistant soils suitable for many types of crops. However, when trees and shrubs are replaced by wheat on the hillsides (i.e., class VI soil), the yields decline progressively and the soil becomes badly eroded. Such major land qualities related to plant growth as availability of water, nutrients, and oxygen; soil texture and depth; salinization and/or alkalinization; possibilities for mechanization; and resistance to erosion are important in determining the suitability of a tract of land for a certain agricultural use.

Vegetational Patterns

The natural vegetation of an ecosystem can be used as an architectural and botanical model for designing and structuring an agroecosystem to replace it. The study of productivity, species composition, efficiency of resource utilization, resistance to pests, leaf area distribution, etc. in natural plant communities is important for building agroecosystems that mimic the structure and function of natural successional ecosystems. In Costa Rica, Ewell et al. (unpublished data) conducted spatial and temporal replacements of wild species by botanically and/or structurally/ecologically similar cultivars. Thus, successional members of the natural system such as Heliconia spp., cucurbitaceous vines, Ipomoea spp., legume vines, shrubs, grasses, and small trees were simulated by plantain, squash varieties, yams, sweet potatoes, local bean crops, Cajanus cajan, corn/sorghum/rice, papaya, cashew, and Cassava spp., respectively. By years two and three, fast-growing tree crops (i.e., Brazil nuts, peach, palm, rosewood, etc.) may form an additional stratum, thus maintaining a continual crop cover, avoiding site degradation and nutrient leaching and providing crop yields throughout the year.

Gast6 and Contreras designed a similar conversion system in the Mediterranean matorral of central Chile. Matorral vegetation consists of shrubs (dominated by Acacia caven) and an understory of mixed grasses. Successful sheep pastures were developed by replacing the natural shrub layer with Atriplex spp. shrubs, a food source for the animals. Thus, species composition was altered, but the structural profile was left intact.

Knowledge of Local Farming Practices

In most rural areas farmers have been cultivating for decades. Some have failed and others have succeeded in developing adapted cropping systems. On small farms in the tropics, for example, farmers have successfully minimized risk and maximized return by intercropping, using low levels of technology and resources. Recent research on polycultures has demonstrated that many characteristics of traditional agroecosystems are more desirable than those of monocultures. Generally, polycultures are more productive, utilize soil resources and photosynthetically active radiation more efficiently, resist pests, epidemics, and weeds better, produce more varied and nutritious food, better utilize local resources and nonhybrid, open-pollinated, locally adapted seeds, and contribute to economic stability, social equality, and farmers' direct participation in decision making. Thus, although tropical small farmers have generally been confined to farming lowquality, marginal soils with little capital or institutional support, their systems provide valuable information for the development of yield-sustaining systems.

The situation is similar for organic farmers in the US and Europe. Recent comparative studies of conventional and organic farming have shown that many organic methods consume less fossil energy, cause less soil erosion, obtain equal profits in most cases, and ensure acceptable yields in the long term. It is not clear whether these yields would be sufficient to meet domestic and export demand.

Reducing Energy use in Food Production

During the last decade, agricultural scientists have become aware that it is important, not only to increase food production, but also to do so with the most efficient use of energy and nonrenewable resources. Some promising approaches to agricultural technology, although valuable, have been based on only one crop production process and have not considered the whole ecosystem.

For the most part, the more integrated approaches are directed toward enhancing photosynthetic efficiency through improvement of plant architecture, use of C4 plants or varieties with a high leafarea index, adoption of efficient planting patterns, and hormonal stimulation of net photosynthesis; improving soil management through minimum tillage, use of living legume mulches, cover cropping, use of manures, enhancement of biological N_2 fixation, and use of mycorrhizae; managing water more efficiently through drip irrigation, mulching, and windbreaks; and managing pests in an ecologically sound manner. These technologies propose minor changes in one or two components of the system, leaving the stringent structure of the monoculture unchallenged, but without doing so, realistic progress cannot be made in the development of sustainable agroecosystems.

However, if the management boundaries are expanded beyond the direct object of control (i.e., a pest problem, soil nutrient deficiency, weed infestation, etc.) a whole new set of management and design options emerge. Of special relevance are those manipulations that can simultaneously affect several components of the system. For example, growers who adopt novel agronomic systems (i.e., multiple cropping or agroforestry systems) can achieve several crop management objectives simultaneously and sometimes require little if any fertilizer or pesticides to sufficiently protect crops and enhance soil fertility. By interplanting wild heliotrope (Heliotropium europaeum) within leguminous crops, weed populations have been reduced about 70% and the abundance of several insect pests reduced below an economic threshold as well. By introducing French and African marigolds in fields of certain crops, populations of nematodes were effectively controlled, and the germination of weeds such as morning glory, pigweed, and Florida beggarweed was also partially inhibited. Adaptations in agriculture along these lines provide a new context for agroecosystem management in which stability of the system depends on manipulating the ecological assemblage in fields to promote biotic interactions that benefit farmers.

Manipulation of Vegetational Diversity

The loss of diversity through the expansion of monocultures has encouraged soil erosion, nutrient depletion, inefficient use of water and energy, reduction of local wildlife, outbreaks of diseases and pests. Restoration of plant diversity through crop rotations, interplantings, agroforestry systems, and cover crops in orchards can correct several of these imbalances. Regional diversification of crop-field boundaries with windbreaks, shelterbelts, and living fences can improve habitat for wildlife and beneficial insects, provide sources of wood, medicinal plants, organic matter, resources for pollinating bees, and, in addition, modify wind speed and the microclimate.

Diversity and Nutrient Cycling

In interplanted agroecosystems the low disturbance and closed canopies promote water and nutrient conservation. Nutrient cycling tends to be rapid. For example, in an agroforestry system, minerals lost by annuals are rapidly taken up by perennial crop plants. In addition, the nutrient robbing propensity of some crops is counteracted by the enriching addition of organic matter to the soil by other crops. Soil nitrogen can be increased by incorporating legumes in the mixture, and phosphorous assimilation can be enhanced somewhat in crops with mycorrhizal associations. In the tropics Ewell et al. found that increased diversity in cropping systems was associated with larger root area, which increases nutrient capture. The maintenance of root systems having large surface areas and an even distribution in the soil profile is desirable for agroecosystems in areas where soil-nutrient storage is often low and leaching rates are high.

Diversity and Plant Diseases

Monocultures are almost invariably prone to diseases. One of the various epidemiological strategies that can be applied to minimize losses due to plant diseases and nematodes is increasing the species and/or genetic diversity of cropping systems. Larios documented evidence of disease buffering in various tropical intercropping schemes. Cowpea intercropped with corn showed less innoculum liberation and dissemination than in cowpea monocultures. The onset of mildew (Oidium manihotis) and scab (Sphaceloma sp.) infestation was delayed on cassava associated with beans and/or sweet potatoes. Cowpea mosaic virus and cowpea chlorotic virus occurred at lower levels in cowpea intercropped with cassava or plantain. The available examples indicate that mixtures of different crop species or varieties (multilines) buffer against disease losses by delaying the onset of the disease, reducing spore dissemination, or modifying microenvironmental conditions such as humidity, light, temperatuie, and air movement. Certain associated plants can function as repellents, antifeedants, growth disrupters, or toxicants. In the case of soilborne pathogens, some plant combinations and organic amendments can enhance soil fungistasis and antibiosis.

Diversity and Weed Populations

The continuous manipulations of the physical environment necessary for modern crop production has favored the selection of opportunistic and highly competitive weeds. Most weed species are stimulated by regular disturbances in monocultures. Of the various factors that influence the crop-weed balance in a field, the density of crop plants and weeds plays a major role in the outcome of competition between them. When the cropping pattern is intensive, the level and type of weed community is a product of the crop and its management. In multiple cropping systems the nature of the crop mixtures (especially canopy closure) can keep the soil covered throughout the growing season, shading out sensitive weed species and minimizing the need for weed control. Intercropping systems of corn/mungbean and corn/ sweet potato are common systems that inhibit weed competition. In these systems the complex canopies with large leaf areas intercept a significant proportion of the incident light, shading out sensitive weed species. In general, weed suppression in intercropping systems depends on the component crops, their density, and the fertility of the soil.

Allelopathy, which is the inhibition of germination, growth, or metabolism of one plant due to the release of organic chemicals by another, is a process that may contribute to increasing the competitiveness of crops over coexisting weeds in mono- and polycultures. Crops such as rye, barley, wheat, tobacco, and oats release toxic substances into the environment, either through root exudation or from decaying plant material, that inhibit the germination and growth of some weed species. Plant leachates from certain varieties of cucumbers have allelopathic effects on prosomillet. Root secretions from rye and oats accessions can inhibit germination and growth of weeds such as wild mustard, Brassica spp., and poppy (Papaver rhoeas).

Diversity and Insect Populations

The exacerbation of most insect-pest problems has been associated with the spatial and temporal expansion of crop monocultures at the expense of the natural vegetation, thereby decreasing local habitat diversity. This simplification can seriously affect the abundance and efficiency of natural enemies, which depend on habitat complexity for sources of alternate prey/hosts, pollen and nectar, shelter, and nesting and overwintering sites.

Plant diversification of agroecosystems can increase environmental opportunities for natural enemies and, consequently, improve biological pest control. Agronomically, there are several ways to design species-rich cropping systems.

Field-margin vegetation and/or withinfield plant diversity can be manipulated by designing mixtures or polycultures of various temporal and spatial arrangements. In Colombia, one of us found that beans grown in dicultures with corn had 25% fewer leafhopper adults (Empoasca kraemeri) than monoculture beans, and population densities of the leafbeetle (Diabrotica balteata) were 45% lower in corn/bean plots than in bean monocultures. The incidence of the fall armyworm (Spodoptera frugiperda) was 23% lower in corn polycultures than in monocultures. Planting times in dicultures can also affect pest interactions. For example, further reductions in leafhopper and fall armyworm densities were achieved by establishing the companion crops 10-20 days before the target crop. The effects of some of these systems on the dynamics of insect populations have recently been discussed by Altieri et al., Bach, and Risch. The effects on insect dynamics of increasing weed diversity by using weedborders and alternate rows of weeds and crops, or by providing weeds in certain periods of the crop growth has been extensively reviewed by Altieri et al., Altieri and Whitcomb, and Cromartie; vegetation-management strategies for natural pest regulation have been discussed by Altieri and Letourneau.

Diversity and Productivity

Commonly, a relative yield advantage is obtained from a polyculture versus a mono-culture. This yield advantage is usually expressed as the "land equivalent ratio" (LER), which expresses the monoculture land area required to produce the same amount as I hectare of polyculture, using the same plant populations. This LER can be expressed as follows:

$$LER = \frac{Px}{Kx} + \frac{Py}{Ky}$$

Where Kx and Ky are the yields per unit area when the crops are grown in monoculture, and Px and Py are the production of the two crop species in a polyculture. If the LER is greater than 1, the polyculture "overyields." Corn-bean dicultures, corn-bean-squash tri-cultures, and most of the agroforestry systems with trees as overstory (for example, cacao and rubber) are examples of "overyielding" polycultures in the Latin American tropics.

Agroecosystems Management

Agroecosystems management integrates economic, ecological and social values to tack-le challenges and find opportunities.

It takes a broad view that ranges from the ground under your feet to your neighboring farms and communities, and from farm to market to consumer:

- To a farmer it means finding a style of farming that pays the bills, that the neigh-bors and community want to support and protect, and that your kids want to continue.

- To a scientist it means operating at the intersection of the agricultural disci-plines, together with farmers. It means considering agriculture as a system, and seeing both problems and opportunities as properties that emerge from the sys-tem rather than one of its parts.

- To a student interested in agriculture, it means the framework for all of your courses and experience, no matter how broad your studies are, and the connec-tion between your focused courses and Ohio agriculture.

- To an environmentalist it means a system of agriculture that enhances environmental qualities like biodiversity and is economically successful as well.

- To a businessperson it means entrepreneurial opportunities that enhance social and environmental bottom lines with equal importance to the economic one.

- To a consumer it means an agriculture that can be trusted to provide healthy food from a healthy ecosystem with a fair return to the farmer.

- To a policy maker it means that long-term simultaneous gains in environmental, social and economic dimensions outweigh short-term gains in any one area.

References

- Agricultural-ecosystem, earth-and-planetary-sciences: sciencedirect.com, Retrieved 25 February, 2019

- Agro-ecosystem-analysis-aesa: ipm-info.org, Retrieved 5 June, 2019

- Agroeco: ucmerced.edu, Retrieved 29 April, 2019

- Bioscience: agroeco.org, Retrieved 9 August, 2019

- What-agroecosystems-management: amp.osu.edu, Retrieved 30 May, 2019

Chapter 4

Ecological Management of Pests and Weeds

There are a variety of ecological methods which can be used to manage pests and weeds such as crop rotation, cover cropping and biological pest control. This chapter has been carefully written to provide an easy understanding of these facets of managing weeds, pests and insects.

Agroecological Concepts to Development of Ecologically based Pest Management Strategies

Most scientists today would agree that conventional modern agriculture faces an environmental crisis. Serious problems such as land degradation, salinization, and pesticide pollution of soil, water, and food chains, depletion of ground water, genetic homogeneity, and associated vulnerability raise serious questions regarding the sustainability of modern agriculture. The causes of the environmental crisis are rooted in the prevalent socio-economic system, which promotes monocultures and the use of high input technologies, and agricultural practices that lead to natural resource degradation. Such degradation is not only an ecological process but also a social, political, and economic process. While productivity issues represent part of the problem of natural resource degradation, addressing the problem of agricultural production must go beyond technological issues and include attention to social, cultural, and economic issues that account for the crisis as well.

The loss of yields due to pests in many crops, despite the substantial increase in the use of pesticides, is a symptom of the environmental crisis affecting agriculture. It is well known that cultivated plants grown in genetically homogeneous monocultures do not possess the necessary ecological defense mechanisms to tolerate pest populations that experience outbreaks. Modern agriculturists have selected crops for high yields and high palatability, making them more susceptible to pests by sacrificing natural resistance for productivity. On the other hand, modern agricultural practices negatively affect pests ' natural enemies, which in turn do not find the necessary environmental resources and opportunities in monocultures to effectively suppress pests. As long as the structure of monocultures is maintained as the structural base of agricultural systems, pest problems will continue to persist. Thus, the major challenge for those advocating ecologically based pest management (EBPM) is to find strategies to overcome the ecological limits imposed by monocultures.

Integrated pest management (IPM) approaches have not addressed the ecological causes of the environmental problems in modern agriculture. There still prevails a narrow view that specific causes affect productivity, and overcoming the limiting factor (e.g., insect pest) via new technologies continues to be the main goal. In many IPM projects the main focus has been to substitute less noxious inputs for the agrochemicals that are blamed for so many of the problems associated with conventional agriculture. Emphasis is now placed on purchased biological inputs such as Bacillus thuringiensis, a microbial pesticide that is now widely applied in place of chemical insecticide. This type of technology pertains to a dominant technical approach called input substitution. The thrust is highly technological, characterized by a limiting factor mentality that has driven conventional agricultural research in the past. Agronomists and other agricultural scientists have for generations been taught the "law of the minimum" as a central dogma. According to this dogma, at any given moment there is a single factor limiting yield increases and that factor can be overcome with an appropriate external input. Once one limiting factor has been surpassed—for example nitrogen deficiency, with urea as the correct input—then yields may rise until another factor, pests for example, becomes the new limiting factor due to increased levels of free nitrogen in the foliage that attracts and supports the herbivore populations. That factor then requires another input— a pesticide in this case—and so on, perpetuating a process of treating symptoms rather than dealing with the real causes that evoked the ecological imbalance.

The addition of biotechnology-based approaches in pest management is merely a new tool to be used as input substitutions to address the problems (e.g., pesticide resistance, pollution, soil degradation) caused by previous agrochemical technologies. Transgenic crops developed for pest control closely follow the paradigm of using a single control mechanism (a pesticide) that, as a strategy, has been shown to fail repeatedly over time against pest insects, pathogens, and weeds. Transgenic crops are likely to increase the use of pesticides and to accelerate the evolution of "super weeds" and resistant insect pests.

The "one gene–one pest" approach emphasized by plant breeders introducing vertical resistance or by biotechnologists developing transgenic crops has proven to be easily overcome by pests that are continuously adapting to new situations and evolving detoxification mechanisms. There are many unanswered ecological questions regarding the impact of the release of transgenic plants and microorganisms into the environment. Among the major environmental risks associated with genetically engineered plants are the unintended transfer to plant relatives of the "transgenes" and the unpredictable ecological effects.

Given the above considerations, agroecological theory predicts that biotechnology will exacerbate the problems of conventional agriculture. By promoting monocultures it will also undermine ecological practices of farming, such as crop rotation and poly-cultures, which are key strategies to break the homogeneous nature of monocultures. As presently conceived, biotechnology does not fit into the broad ideals of sustainable agriculture.

This production-oriented viewpoint has diverted agriculturists from realizing that lim-iting factors only represent symptoms of a more systematic disease inherent to imbal-ances within the agroecosystem. This viewpoint has also diverted them from an appre-ciation of the complexity of agroecological processes, thus underestimating the root causes of agricultural limitations. A useful framework to achieve a deeper knowledge of the dynamics of agroecosystems is to use agroecological principles. Agroecology goes beyond a one-dimensional view of agroecosystems and includes their genetics, agron-omy, and edaphology in order to embrace an understanding of ecological and social levels of coevolution, structure, and function. For agroecologists, sustainable yield in the agroecosystem derives from the proper balance of crops, soils, nutrients, sunlight, moisture, and other coexisting organisms. The agroecosystem is productive and healthy when these balanced and rich growing conditions prevail and when crop plants remain resilient to tolerate stress and adversity. Occasional disturbances can be overcome by a vigorous agroecosystem that is adaptable and diverse enough to recover once the stress has passed. Occasionally strong measures (e.g., microbial insecticides, alternative fer-tilizers) may need to be applied by farmers employing alternative methods to control specific pests or soil problems. Agroecology provides the guidelines to carefully manage agroecosystems without unnecessary or irreparable damage. Simultaneous with the struggle to fight pests or diseases, the agroecologist strives to restore the resiliency and health of the agroecosystem. If the cause of disease or pests and so forth is recognized as an imbalance, then the goal of the agroecological treatment is to recover the balance. In agroecology, biodiversification is the primary technique to evoke self-regulation and sustainability.

From a management perspective, the agroecological objective is to provide a balanced environment, sustained yields, biologically mediated soil fertility, and natural pest regulation through the design of diversified agroecosystems and the use of low-input technologies. The strategy is based on ecological principles that lead crop manage-ment to optimal recycling of nutrients and organic matter turnover, closed energy flows, water and soil conservation, and a balance between pest and natural enemy populations. The strategy exploits the complementary and synergistic attributes that result from the various combinations of crops, trees, and animals in spatial and tem-poral arrangements. These combinations determine the establishment of a planned and associated functional biodiversity, which performs key ecological services in the agroecosystem.

The optimal behavior of agroecosystems depends on the level of interactions between and among the various biotic and abiotic components. By assembling a functional bio-diversity, it is possible to initiate synergistic responses that subsidize agroecosystem processes by providing ecological services, such as the activation of soil biology, the recycling of nutrients, the enhancement of beneficial arthropods and antagonists, and so on.

In other words, ecological concepts are utilized to favor natural processes and biological interactions that optimize synergies so that diversified farms are able to sponsor their own soil fertility, crop protection, and productivity. By assembling crops, animals, trees, soils, and other factors in spatially and or temporally diversified schemes, several processes are optimized. Such processes (such as organic matter accumulation, nutrient cycling, natural control mechanisms, etc.) are crucial in determining the sustainability of agricultural systems.

Agroecology takes greater advantage of natural processes and beneficial on-farm interactions in order to reduce off-farm input use and to improve the efficiency of farming systems. Technologies emphasized tend to enhance the functional biodiversity of agroecosystems as well as the conservation of existing on-farm resources. Promoted technologies are multifunctional, as their adoption usually means favorable changes in various components of the farming systems at the same time.

For example, legume-based crop rotations are one of the simplest forms of biodiversification and can simultaneously optimize soil fertility and pest regulation. It is well known that rotations improve yields by the known action of interrupting weed, disease, and insect life cycles. However, they can also have subtle effects such as enhancing the growth and activity of soil organisms, including vesicular arbuscular mycorrhizae, which allow crops to more efficiently use soil nutrients and water.

Another practice is cover cropping or the growing of pure or mixed stands of legumes and cereals to protect the soil against erosion, which ameliorates soil structure, enhances soil fertility, and suppresses pests including weeds, insects, and pathogens. Cover crops can improve soil structure and water penetration, prevent soil erosion, modify the microclimate, and reduce weed competition. Besides these effects, cover crops can affect the dynamics of orchards and vineyards by enhancing soil biology and fertility and by increasing the biological control of insect pest populations through the harboring of predators and parasites.

Perhaps the most dramatic example of the integrative effects of a multipurpose technology in simultaneously enhancing IPM and soil fertility is organic soil management. Some studies suggest the physiological susceptibility of crops to insects is affected by the form of fertilizer used (organic vs. chemical fertilizer). Studies documenting lower density of several insect herbivores in low-input farming systems have partly attributed such reduction to lower nitrogen content in the organically farmed crops.

The ultimate goal of agroecological design is to integrate components so that overall biological efficiency is improved, biodiversity is preserved, and the agroecosystem productivity and its self-sustaining capacity is maintained. The goal is to design an agroecosystem that mimics the structure and function of natural ecosystems. These systems typically include:

- Vegetative cover as an effective soil- and water-conserving measure, met through the use of no-till practices, mulch farming, and cover crops and other appropriate methods;

- Regular supply of organic matter through the addition of organic matter (manure and compost) that results in the promotion of soil biotic activity;

- Nutrient recycling mechanisms through the use of crop rotations and crop/livestock systems based on the use of legumes; and

- Pest regulation assured through enhanced activity of biological control agents achieved by introducing and/or conserving natural enemies and antagonists.

The process of converting a conventional crop production system that relies heavily on systemic, petroleum-based inputs to a diversified agroecosystem with low inputs is not simply a process of withdrawing external inputs without compensatory replacement or alternative management. Considerable ecological knowledge is required to direct the array of natural flows necessary to sustain yields in a low-input system. The process of conversion from a high-input conventional management to a low-external-input management is a transitional process with four marked phases:

- Progressive chemical withdrawal;

- Rationalization and efficiency of agrochemical use through integrated pest management and integrated nutrient management;

- Input substitution—using alternative, low-energy input technologies; and

- Redesign of diversified farming systems with an optimal crop/animal integration, which encourages synergism so that the system can sponsor its own soil fertility, natural pest regulation, and crop productivity.

During the four phases, management is guided to ensure the following processes:

- Increasing biodiversity both in the soil and above ground;

- Increasing biomass production and soil organic matter content;

- Decreasing levels of pesticide residues and losses of nutrients and water components;

- Establishment of functional relationships between the various plant and animal components on the farm; and

- Optimal planning of crop sequences and combinations and efficient use of locally available resources.

The challenge for EBPM scientists is to identify the correct management techniques and crop assemblages that, through their biological synergism, will provide key ecological services that sustain the performance of agroecosystems. The exploitation of these synergisms in real farm settings involves agroecosystem design and management that require an understanding of the numerous relationships among soils, plants, herbivores, and natural enemies. Clearly, the emphasis of this approach is to restore natural control mechanisms through the addition of selective biodiversity components within and outside the crop field, thereby creating a whole array of possible arrangements of vegetation in time and space.

Biological Pest Control

Biological control, biocontrol, or biological pest control is a method of suppressing or controlling the population of undesirable insects, other animals, or plants by the introduction, encouragement, or artificial increase of their natural enemies to economically non–important levels. It is an important component of integrated pest management (IPM) programs.

The biological control of pests and weeds relies on predation, parasitism, herbivory, or other natural mechanisms. Therefore, it is the active manipulation of natural phenomena in serving human purpose, working harmoniously with nature. A successful story of biological control of pests refer to the human beings' capability to depict natural processes for their use and can be the most harmless, non–polluting, and self–perpetuating control method.

In biological control, the reduction of pest populations is achieved by actively using natural enemies.

Predatory Polistes wasp looking for bollworms or other caterpillarson a cotton plant.

Natural enemies of the pests, also known as biological control agents, include predatory and parasitoidal insects, predatory vertebrates, nematode parasites, protozoan parasites, and fungal, bacterial, as well as viral pathogens. Biological control agents of plant diseases are most often referred to as antagonists. Biological control agents of weeds include herbivores and plant pathogens. Predators, such as lady beetles and lacewings, are mainly free–living species that consume a large number of prey during their lifetime. Parasitoids are species whose immature stage develops on or within a single insect host, ultimately killing the host. Most have a very narrow host range. Many species of wasps and some flies are parasitoids. Pathogens are disease–causing organisms including bacteria, fungi, and viruses. They kill or debilitate their host and are relatively specific to certain pest or weed groups.

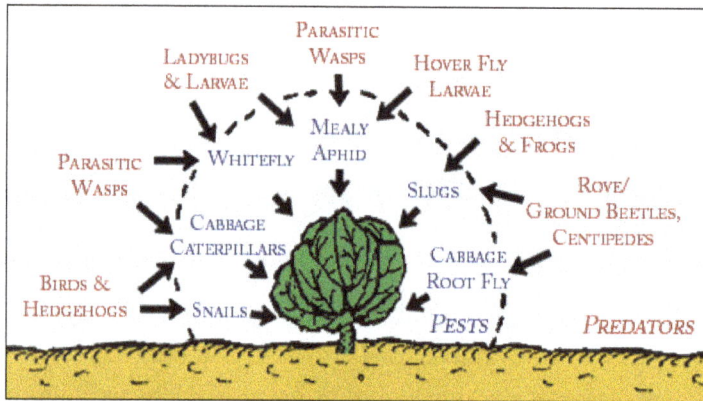

Diagram illustrating the natural enemies of cabbage pests.

Strategies of Biological Control Methods

There are three basic types of biological control strategies; conservation biocontrol, classical biological control, and augmentative biological control (biopesticides).

Conservation Biocontrol

The conservation of existing natural enemies is probably the most important and readily available biological control practice available to homeowners and gardeners. Natural enemies occur in all areas, from the backyard garden to the commercial field. They are adapted to the local environment and to the target pest, and their conservation is generally simple and cost–effective. For example, snakes consume a lot or rodent and insect pests that can be damaging to agricultural crops or spread disease. Dragonflies are important consumers of mosquitoes.

Eggs, larvae, and pupae of Helicoverpa moths, the main insect pests of cotton, are all attacked by many beneficial insects and research can be conducted in identifying critical habitats, resources needed to maintain them, and ways of encouraging their activity. Lacewings, lady beetles, hover fly larvae, and parasitized aphid mummies are almost always present in aphid colonies. Fungus–infected adult flies are often common following periods of high humidity. These naturally occurring biological controls are often susceptible to the same pesticides used to target their hosts. Preventing the accidental eradication of natural enemies is termed simple conservation.

Classical Biological Control

Classical biological control is the introduction of exotic natural enemies to a new locale where they did not originate or do not occur naturally. This is usually done by government authorities.

In many instances, the complex of natural enemies associated with an insect pest may be inadequate. This is especially evident when an insect pest is accidentally introduced into a new geographic area without its associated natural enemies. These introduced pests are referred to as exotic pests and comprise about 40 percent of the insect pests in the United States. Examples of introduced vegetable pests include the European corn borer, one of the most destructive insects in North America.

To obtain the needed natural enemies, scientists have utilized classical biological control. This is the practice of importing, and releasing for establishment, natural enemies to control an introduced (exotic) pest, although it is also practiced against native insect pests. The first step in the process is to determine the origin of the introduced pest and then collect appropriate natural enemies associated with the pest or closely related species. The natural enemy is then passed through a rigorous quarantine process, to ensure that no unwanted organisms (such as hyperparasitoids or parasites of the parasite) are introduced, and then they are mass produced, and released. Follow–up studies are conducted to determine if the natural enemy becomes successfully established at the site of release, and to assess the long–term benefit of its presence.

There are many examples of successful classical biological control programs. One of the earliest successes was with the cottony cushion scale (Icerya purchasi), a pest that was devastating the California citrus industry in the late 1800s. A predatory insect, the Australian lady beetle or vedalia beetle (Rodolia cardinalis), and a parasitoid fly were introduced from Australia. Within a few years, the cottony cushion scale was completely controlled by these introduced natural enemies. Damage from the alfalfa weevil, a serious introduced pest of forage, was substantially reduced by the introduction of several natural enemies like imported ichnemonid parasitoid Bathyplectes curculionis. About twenty years after their introduction, the alfalfa area treated for alfalfa weevil in the northeastern United States was reduced by 75 percent. A small wasp, Trichogramma ostriniae, introduced from China to help control the European corn borer (Pyrausta nubilalis), is a recent example of a long history of classical biological control efforts for this major pest. Many classical biological control programs for insect pests and weeds are under way across the United States and Canada.

Classical biological control is long lasting and inexpensive. Other than the initial costs of collection, importation, and rearing, little expense is incurred. When a natural enemy is successfully established it rarely requires additional input and it continues to kill the pest with no direct help from humans and at no cost. Unfortunately, classical biological control does not always work. It is usually most effective against exotic pests and less so against native insect pests. The reasons for failure are often not known, but may include the release of too few individuals, poor adaptation of the natural enemy to environmental conditions at the release location, and lack of synchrony between the life cycle of the natural enemy and host pest.

Augmentative Biological Control

This third strategy of biological control method involves the supplemental release of natural enemies. Relatively few natural enemies may be released at a critical time of the season (inoculative release) or literally millions may be released (inundative release). Additionally, the cropping system may be modified to favor or augment the natural enemies. This latter practice is frequently referred to as habitat manipulation.

An example of inoculative release occurs in greenhouse production of several crops. Periodic releases of the parasitoid, Encarsia formosa, are used to control greenhouse whitefly, and the predaceous mite, Phytoseilus persimilis, is used for control of the two–spotted spider mite. The wasp Encarsia formosa lays its eggs in young whitefly "scales," turning them black as the parasite larvae pupates. Ideally it is introduced as soon as possible after the first adult whitefly are seen. It is most effective when dealing with low level infestations, giving protection over a long period of time. The predatory mite, Phytoseilus persimilis, is slightly larger than its prey and has an orange body. It develops from egg to adult twice as fast as the red spider mite and once established quickly overcomes infestation.

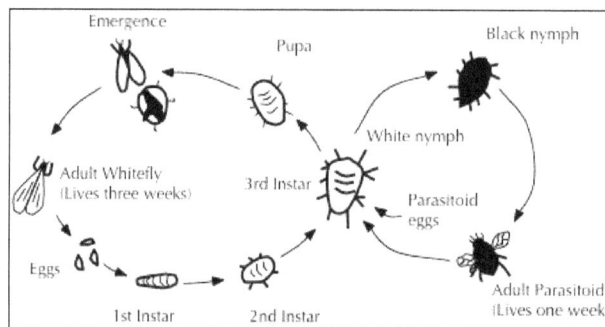

Diagram illustrating the life cycles of Greenhouse whitefly and its parasitoid wasp Encarsia Formosa.

Lady beetles, lacewings, or parasitoids such as Trichogramma are frequently released in large numbers (inundative release) and are often known as biopesticides. Recommended release rates for Trichogramma in vegetable or field crops range from 5,000 to 200,000 per acre per week depending on level of pest infestation. Similarly, entomoparasitic nematodes are released at rates of millions and even billions per acre for control of certain soil-dwelling insect pests. Entomopathogenic fungus Metarhizium anisopliae var. acridum, which is specific to species of short–horned grasshoppers

(Acridoidea and Pyrgomorphoidea) widely distributed in Africa, has been developed as inundative biological control agent.

Habitat or environmental manipulation is another form of augmentation. This tactic involves altering the cropping system to augment or enhance the effectiveness of a natural enemy. Many adult parasitoids and predators benefit from sources of nectar and the protection provided by refuges such as hedgerows, cover crops, and weedy borders. Mixed plantings and the provision of flowering borders can increase the diversity of habitats and provide shelter and alternative food sources. They are easily incorporated into home gardens and even small-scale commercial plantings, but are more difficult to accommodate in large–scale crop production. There may also be some conflict with pest control for the large producer because of the difficulty of targeting the pest species and the use of refuges by the pest insects as well as natural enemies.

Examples of habitat manipulation include growing flowering plants (pollen and nectar sources) near crops to attract and maintain populations of natural enemies. For example, hover fly adults can be attracted to umbelliferous plants in bloom.

Biological control experts in California have demonstrated that planting prune trees in grape vineyards provides an improved overwintering habitat or refuge for a key grape pest parasitoid. The prune trees harbor an alternate host for the parasitoid, which could previously overwinter only at great distances from most vineyards. Caution should be used with this tactic because some plants attractive to natural enemies may also be hosts for certain plant diseases, especially plant viruses that could be vectored by insect pests to the crop. Although the tactic appears to hold much promise, only a few examples have been adequately researched and developed.

Different Types of Biological Control Agents

Predators

Ladybird larva eating wooly apple aphids

Ladybugs, and in particular their larvae which are active between May and July in the northern hemisphere, are voracious predators of aphids such as greenfly and blackfly, and will also consume mites, scale insects, and small caterpillars. The ladybug is a very familiar beetle with various colored markings, while its larvae are initially small and spidery, growing up to 17 millimeters (mm) long. The larvae have a tapering segmented gray/black body with orange/yellow markings nettles in the garden and by leaving hollow stems and some plant debris over–winter so that they can hibernate over winter.

Hoverflies, resembling slightly darker bees or wasps, have characteristic hovering, darting flight patterns. There are over 100 species of hoverfly, whose larvae principally feed upon greenfly, one larva devouring up to 50 a day, or 1000 in its lifetime. They also eat fruit tree spider mites and small caterpillars. Adults feed on nectar and pollen, which they require for egg production. Eggs are minute (1 mm), pale yellow-white, and laid singly near greenfly colonies. Larvae are 8–17 mm long, disguised to resemble bird droppings; they are legless and have no distinct head. Therefore, they are semi–transparent with a range of colors from green, white, brown, and black. Hoverflies can be encouraged by growing attractant flowers such as the poached eggplant (Limnanthes douglasii), marigolds, or phacelia throughout the growing season.

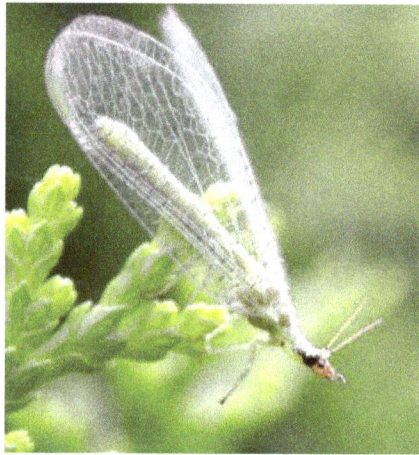

Lacewings are available from biocontrol dealers.

Dragonflies are important predators of mosquitoes, both in the water, where the dragonfly naiads eat mosquito larvae, and in the air, where adult dragonflies capture and eat adult mosquitoes. Community–wide mosquito control programs that spray adult mosquitoes also kill dragonflies, thus removing an important biocontrol agent, and can actually increase mosquito populations in the long term.

Other useful garden predators include lacewings, pirate bugs, rove and ground beetles, aphid midge, centipedes, as well as larger fauna such as frogs, toads, lizards, hedgehogs, slow–worms, and birds. Cats and rat terriers kill field mice, rats, june bugs, and birds. Dogs chase away many types of pest animals. Dachshunds are bred specifically to fit inside tunnels underground to kill badgers.

Parasitoidal Insects

Most insect parasitoids are wasps or flies. For example, the parasitoid Gonatocerus ashmeadi (Hymenoptera: Mymaridae) has been introduced to control the glassy-winged sharpshooter Homalodisca vitripennis (Hemipterae: Cicadellidae) in French Polynesia and has successfully controlled about 95 percent of the pest density. Parasitiods comprise a diverse range of insects that lay their eggs on or in the body of an insect host, which is then used as a food for developing larvae. Parasitic wasps take much longer than predators to consume their victims, for if the larvae were to eat too fast they would run out of food before they became adults. Such parasites are very useful in the organic garden, for they are very efficient hunters, always at work searching for pest invaders. As adults, they require high–energy fuel as they fly from place to place, and feed upon nectar, pollen and sap, therefore planting plenty of flowering plants, particularly buckwheat, umbellifers, and composites will encourage their presence.

Four of the most important groups are:

- Ichneumonid wasps: (5–10 mm) Prey mainly on caterpillars of butterflies and moths.

- Braconid wasps: Tiny wasps (up to 5 mm) attack caterpillars and a wide range of other insects including greenfly. It is a common parasite of the cabbage white caterpillar, seen as clusters of sulphur yellow cocoons bursting from collapsed caterpillar skin.

- Chalcid wasps: Among the smallest of insects (<3 mm). It parasitizes eggs/larvae of greenfly, whitefly, cabbage caterpillars, scale insects, and strawberry tortrix moth.

- Tachinid flies: Parasitize a wide range of insects including caterpillars, adult and larval beetles, true bugs, and others.

Parasitic Nematodes

Nine families of nematodes (Allantone-matidae, Diplogasteridae, Heterorhabditidae, Mermithidae, Neotylenchidae, Rhabditidae, Sphaerulariidae, Steinernematidae, and Tetradonematidae) include species that attack insects and kill or sterilize them, or alter their development. In addition to insects, nematodes can parasitize spiders, leeches, (annelids), crustaceans and mollusks. An excellent example of a situation in which a nematode may replace chemicals for control of an insect is the black vine weevil, Otiorhynchus sulcatus, in cranberries. Uses of chemical insecticides on cranberry either are restricted or have not provided adequate control of black vine weevil larvae. Heterorhabditis bacteriophora NC strain was applied, and it provided more than 70 percent control soon after treatment and was still providing that same level of control a year later.

Many nematode–based products are currently available. They are formulated from various species of Steinernema and Heterorhabditis. Some of the products found in

various countries are ORTHO Bio–Safe, BioVector, Sanoplant, Boden-Ntitzlinge, He-
lix, Otinem, Nemasys, and so forth (Smart 1995). A fairly recent development in the
control of slugs is the introduction of "Nemaslug," a microscopic nematode (Phasmar-
habditis hermaphrodita) that will seek out and parasitize slugs, reproducing inside
them and killing them. The nematode is applied by watering onto moist soil, and gives
protection for up to six weeks in optimum conditions, though is mainly effective with
small and young slugs under the soil surface.

Plants to Regulate Insect Pests

Choosing a diverse range of plants for the garden can help to regulate pests in a variety
of ways, including:

- Masking the crop plants from pests, depending on the proximity of the companion or intercrop.
- Producing olfactory inhibitors, odors that confuse and deter pests.
- Acting as trap plants by providing an alluring food that entices pests away from
  crops.
- Serving as nursery plants, providing breeding grounds for beneficial insects.
- Providing an alternative habitat, usually in a form of a shelterbelt, hedgerow, or
  beetle bank, where beneficial insects can live and reproduce. Nectar–rich plants
  that bloom for long periods are especially good, as many beneficials are nectivo-
  rous during the adult stage, but parasitic or predatory as larvae. A good example
  of this is the soldier beetle, which is frequently found on flowers as an adult, but
  whose larvae eat aphids, caterpillars, grasshopper eggs, and other beetles.

The following are plants often used in vegetable gardens to deter insects:

Plant	Pests
Basil	Repels flies and mosquitoes.
Catnip	Deters flea beetle.
Garlic	Deters Japanese beetle.
Horseradish	Deters potato bugs.
Marigold	The workhorse of pest deterrents. Discourages Mexican bean beetles, nema-todes and others.
Mint	Deters white cabbage moth, ants.
Nasturtium	Deters aphids, squash bugs and striped pumpkin beetles.
Pot Marigold	Deters asparagus beetles, tomato worm, and general garden pests.
Peppermint	Repels the white cabbage butterfly.
Rosemary	Deters cabbage moth, bean beetles and carrot fly.
Sage	Deters cabbage moth and carrot fly.
Southernwood	Deters cabbage moth.
Summer Savory	Deters bean beetles.

Tansy	Deters flying insects, Japanese beetles, striped cucumber beetles, squash bugs and ants.
Thyme	Deters cabbage worm.
Wormwood	Deters animals from garden.

Pathogens to be used as Biopesticides

Various bacterial species are widely used in controlling the pests as well as weeds. The best–known bacterial biological control which can be introduced in order to control butterfly caterpillars is Bacillus thuringiensis, popularly called Bt. This is available in sachets of dried spores, which are mixed with water and sprayed onto vulnerable plants such as brassicas and fruit trees. After ingestion of the bacterial preparation, the endotoxin liberated and activated in the midgut will kill the caterpillars, but leave other insects unharmed. There are strains of Bt that are effective against other insect larvae. Bt. israelensis is effective against mosquito larvae and some midges.

Viruses most frequently considered for the control of insects (usually sawflies and Lepidoptera) are the occluded viruses, namely NPV, cytoplasmic polyhedrosis (CPV), granulosis (GV), and entomopox viruses (EPN). They do not infect vertebrates, non–arthropod invertebrates, microorganisms, and plants. The commercial use of virus insecticides has been limited by their high specificity and slow action.

Fungi are pathogenic agents to various organisms including the pests and the weeds. This feature is intensively used in biocontrol. The entomopathogenic fungi, like Metarhizium anisopliae, Beauveria bassiana, and so forth cause death to the host by the secretion of toxins. A biological control being developed for use in the treatment of plant disease is the fungus Trichoderma viride. This has been used against Dutch Elm disease, and to treat the spread of fungal and bacterial growth on tree wounds. It may also have potential as a means of combating silver leaf disease.

Significance of Biological Control

Biological control proves to be very successful economically, and even when the method has been less successful, it still produces a benefit–to–cost ratio of 11:1. The benefit–to–cost ratios for several successful biological controls have been found to range from 1:1 to 250:1. Further, net economic advantage for biological control without scouting vs. conventional insecticide control ranged from $ 7.43 to $ 0.12 per hectare in some places. It means that even if the yield produce under biological control be below that for insecticidal control by as much as 29.3 kilos per hectare, the biological control would not lose its economic advantage.

Biological control agents are non–polluting and thus environmentally safe and acceptable. Usually they are species specific to targeted pest and weeds. The biological control discourages the use of environmentally and ecologically unsuitable chemicals, so it always leads to the establishment of natural balance. The problems of increased resistance in the pest will not arise, as both biological control agents and the pests are in complex race of evolutionary dynamism. Because of chemical resistance developed by the Colorado potato beetle (CPB), its control has been achieved by the use of bugs and beetles (Hein).

Negative Results of Biological Control

Biological control tends to be naturally self–regulating, but as ecosystems are so complex, it is difficult to predict all the consequences of introducing a biological controlling agent (HP 2007). In some cases, biological pest control can have unforeseen negative results, that could outweigh all benefits. For example, when the mongoose was introduced to Hawaii in order to control the rat population, it predated on the endemic birds of Hawaii, especially their eggs, more often than it ate the rats. Similarly, the introduction of the cane toad to Australia 50 years ago to eradicate a beetle that was destroying sugar beet has been spreading as a pest throughout eastern and northern Australia at a rate of 35 km/22 mi a year. Since the cane toad is poisonous, it has few Australian predators to control its population.

Weed Ecology

Weed ecology might be described as the weed's lifestyle - where does it tend to thrive, what are its particular needs and adaptations to its environment? People have a major influence on weed ecology such as weed transport/spread and creating various cropping and management techniques that favor a particular weed species. Also, weedy plants have the capability to survive under adverse conditions and use adaptive mechanisms that enable them to thrive under conditions of extreme soil disturbance. Understanding weed ecology can help in designing management systems to minimize their impact.

Weed Biology

Effective weed management is dependent upon the biology of the weed including type and lifecycle. Weeds are more easily identified and/or managed at certain times during their lifecycle.

Life Cycles

Summer Annuals

Seeds germinate in spring, vegetative during spring/summer, and flowers and sets seed in mid to late summer, die in fall.

Examples:

- Common lambsquarters
- Pigweed
- Hairy galinsoga
- Velvetleaf
- Large crabgrass
- Common ragweed
- Giant foxtail
- Common purslane
- Smartweed

Winter Annuals

Seeds germinate in fall and winter, vegetative during spring, and flowers and set seed in late spring/early summer before it dies.

Examples:

- Common chickweed
- Owny brome
- Shepherdspurse
- Henbit

Biennials

Plants live more than one year over two growing seasons; seeds germinate and over-winter in a rosette stage, then complete life the following year.

Examples:

- Wild carrot
- Common teasel
- Bull thistle
- Common burdock

Perennials

Simple: generally spread by seed; large root but no spreading horizontal roots.

Examples:

- Dandelion
- Common pokeweed
- Dock
- Plaintain

Creeping: reproduce by rhizomes and stolons; some seed reproduction.

Examples:

- Canada thistle
- Poison ivy
- Multiflora rose
- Quackgrass
- Ground ivy
- Hemp dogbane
- Yellow nutsedge
- Japanese knotweed
- Groundcherry

Weed Identification

Effective weed management requires proper identification of weeds - otherwise tactics chosen may not work and end up costing money and frustration. Below are some of the key features to look at when identifying a plant accused of being a weed! Many good books, fact sheets and websites now feature photo galleries of weeds at various stages to help with identification.

Integrated Weed Management

Integrated Weed Management (IWM) is an approach to managing weeds using multiple control tactics. The purpose of IWM is to include many methods in a growing season to allow producers the best chance to control troublesome weeds.

Need for IWM

It might be better to first discuss why weed control is necessary. Weeds negatively impact crop yields, interfere with many crop production practices, and weed seeds can contaminate grain. Based on national research, corn and soybean yield can be reduced by approximately 50% without effective weed control.

Herbicide application is the main weed control strategy used. Reliance on this one method has led to the development of herbicide-resistant weeds. There are a limited number of herbicides available to use and cases of herbicide resistance are rapidly increasing in the US. As a result, herbicides are in need of extra help to continue to ensure adequate weed control.

It is imperative to integrate non-herbicide weed management tactics now to control weeds rather than relying on the ag-chemical industry to continue to develop new herbicides.

Components of an IWM Plan

The goal of IWM is to incorporate different methods of weed management into a combined effort to control weeds. Just as using the same herbicide again and again can lead to resistance, reliance on any one of the methods below over time can reduce its efficacy against weeds. Two major factors to consider when developing an IWM plan are (1) target weed species and (2) time, resources, and capabilities necessary to implement these tactics.

While it is wise to be a good steward of herbicide technology, through the use of PRE and POST herbicide applications or tank mixes, IWM requires the use of tactics beyond herbicides. For example, using these herbicide application practices along with a winter cover crop or harvest weed seed control (HWSC) and prevention methods would be considered IWM.

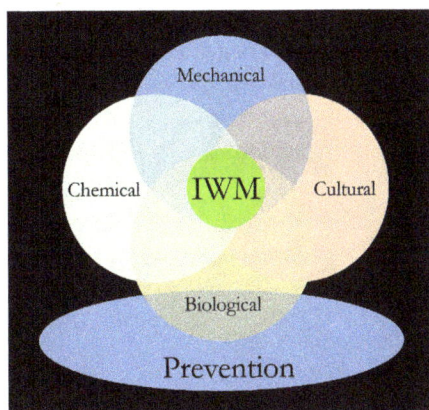

IWM is composed of mechanical, cultural, chemical and biological tactics.

Categories of IWM Practices

Prevention

Prevention is one of the first steps of weed management. This category is unlike the others in that it focuses on keeping weeds out of the field or spreading within a field.

Growers can incorporate this tactic by:

- Avoiding crop seed, manure, and other inputs that are contaminated with weed seeds.

- Cleaning equipment, including combines (combine cleaning methodology), that could transport weed seeds between fields.

- Preventing weeds from producing seeds in the field but also in ditches, fence-rows, and other nearby non-crop areas.

- Scouting for weeds in a timely manner.

- Proceeding with caution when purchasing used farm equipment or using rented land.

Horseweed seed on bush hog after mowing a weedy field.
These seeds can easily spread to other fields.

Feeding combine with straw bales for deep cleaning.

Cultural

A healthy, vigorous crop is the best weed control. Cultural practices are designed to give the crop a competitive advantage over weeds.

Growers can incorporate this tactic by:

- Reduced row spacing so the crop can reach canopy more quickly to shade out weeds.

- Crop rotation to prevent weeds from adapting to the weed control tactics common in any one crop.

- Nutrient management to allow optimum crop uptake while denying weeds access to nutrients.

- Cover crops to compete with weeds for space, sunlight, nutrients, and water.

- Altered planting dates to give the crop a head start or allow for a flush of weed germination that can be controlled before planting.

- Crop variety selection to ensure crops have the utmost competitive advantage against weeds.

Horseweed suppression from a cover crop mixture compared to an area where no cover crops were planted.

Chemical

Herbicides are an integral part of most weed management plans and will continue to be so, even in IWM programs.

Good management practices for applying herbicides include:

- Timely scouting.

- Proper weed identification and awareness of what herbicide-resistant weeds are in the area.

- Correct herbicide application, meaning applying the appropriate product at the right rate and at the right time.

- Maximized diversity through the use of tank mixes herbicides with different, effective sites of action (SOA) and by rotating herbicides throughout the season whenever possible.

- Plan ahead across seasons to avoid using herbicides with the same SOA repeatedly.

•

Understanding the concept of herbicide Site of Action (SOA) is key to effectively managing herbicide resistance.

Mechanical

Mechanical weed management focuses on physical practices that disrupt germination and destroy plant tissue.

Growers can incorporate this tactic by:

- Hand-pulling,
- Tillage,
- Burning,
- Mowing,
- Robotic weeding machines,
- Harvest weed seed control, which reduces the input of weed seeds into the soil seedbank by destroying or removing seeds retained on the weeds at the time of harvest.

Hand pulling escaped weeds is critical to prevent seeds from entering the soil seed bank, particularly for herbicide resistant weeds such as Palmer amaranth.

Windrow burning, a form of harvest weed seed control, is an excellent tactic to prevent weed seeds from entering the soil seed bank.

Harrington Seed Destructor: Two mills destroy weed seeds contained in the chaff portion that comes out from the combine.

Biological

This tactic uses living organisms to target weeds including bacteria, fungi, or insects that have a preference for a certain weed species. This tactic is arguably the least used of all tactics but is the subject of much research. Cover crops can be considered a biological control tactic.

References

- Biological-pest-control: newworldencyclopedia.org, Retrieved 8 July, 2019

- Weed-ecology-biology-and-identification: extension.psu.edu, Retrieved 28 February, 2019

- What-is-integrated-weed-management: integratedweedmanagement.org, Retrieved 18 May, 2019

- Steps-toward-ecological-weed-management-in-organic-vegetables: extension.org, Retrieved 9 January, 2019

- Sustainable-grazing, sustainable-agriculture, agriculture: acs.edu.au, Retrieved 19 March, 2019

Chapter 5

Plant Diseases: Ecology and Management

Plant diseases are an integral part of the function of natural ecosystems. They are responsible for the maintenance of plant populations as well as their composition and diversity. The chapter closely examines the key concepts of plant disease ecology and ecological disease management to provide an extensive understanding of the subject.

Ecology of Plant Diseases and other Microorganism-plant Interactions

Every plant in a tropical forest interacts intimately with microorganisms from the time it germinates as a seed until the mature plant dies and returns to the soil. An appreciation of the ecology of these interactions is essential for understanding how tropical forests work as well as for effective forest management. Microorganisms can benefit the host plant through improved nutrition and defense against pests, or they can cause deleterious effects including reduced growth and reproduction, or even death. Decomposition of plant material by microbes is critical to nutrient cycling. Although plant diseases may have negative effects on individual plants, disease should be viewed as an important, natural part of the function of any natural ecosystem. Plant diseases can play key roles in regulating the dynamics and distributions of plant populations, determining the composition and diversity of plant communities, facilitating successional processes, providing wildlife habitat, and determining the success of restoration efforts.

Kinds of Plant-microbe Interactions

Plants and microorganisms interact in many ways. The simplest interaction is that of fungi and bacteria decomposing dead plant material. Fungi are particularly important in the decomposition of wood, where extracellular enzymes break down complex organic molecules such as lignin and cellulose into ever-simpler molecules, converting highly structured plant material into soil, and making available trapped nutrients for plant uptake.

Much more complex are the interactions between microbes and living plants. The microbe and plant enter into a symbiosis (living together) where the plant is called the host and the microbe the symbiont. Symbioses may take many forms. The symbiont may obtain its nutrition from the host to the host's detriment, a type of symbiosis termed parasitism. Alternatively, the symbiont may live in the host but without causing significant negative effects: a relationship called commensalism. Finally, mutualism is a symbiosis where both the symbiont and the host benefit from the relationship. Unfortunately, particular plant-microbe interactions are rarely easy to categorize because these three types of symbioses fall along a continuum. As we will see below, the confusion increases further when a microbe is a commensal or even a mutualist at one stage in its life cycle, and a parasite at another - even on the same host.

A plant is diseased when a persistent agent (usually a parasite) disrupts the normal functions or form of the host plant, leading to impairment or death of the plant or parts of the plant. Disease is differentiated from injury by the persistence of association between the disease agent and the host (thus a machete cut is not a disease), and from simple parasitism in that the effect on the host are more extensive or damaging than would be expected from the simple removal of nutrients or water from the host. When a parasite causes a disease on the host, it is called a pathogen.

For all types of symbioses the association may be obligate (required for survival for the symbiont, the host, or both) or facultative (optional for either or both). Obligate pathogens can grow and reproduce only on a living host plant. Other pathogens also have saprotrophic phases (consuming dead organic matter), and only facultatively infect and cause disease on plants. Opportunistic pathogens facultatively attack plants that are already sick or dying from severe stress or damage from other pathogens. Some may cause little damage to the plant, but act almost as a saprotroph, primarily consuming tissue killed by an earlier pathogen.

Natural History of Diseases

The Disease Triangle

The nature of a symbiosis depends on three interacting factors: the host, the symbiont, and the environment. Whereas the genetic makeup of both the plant and the microbe are critical in determining the range of possible outcomes of a symbiosis, the environment influences what the final outcome of the interaction will be. This tripartite phenomenon is traditionally discussed within the context of plant diseases, but is easily extended to any plantmicrobe interaction. In the case of a plant disease, a virulent pathogen must come in contact with a susceptible host in order for disease to develop. However, even if spores of a pathogenic fungus do land on a susceptible host plant, if the environmental conditions are too dry, too wet, too hot, or too cold for the spores to germinate, infection will not take place and disease will not develop, although there is potential for disease development under the appropriate environmental conditions. Similarly, a fungus might infect but not cause disease in a host plant growing vigorously in moist, fertile soil in adequate sunlight, but the same fungus may kill a host plant should the interaction take place under more stressful (to the plant) conditions. Changes in the environment during tenure of a symbiosis can change the nature of the interaction.

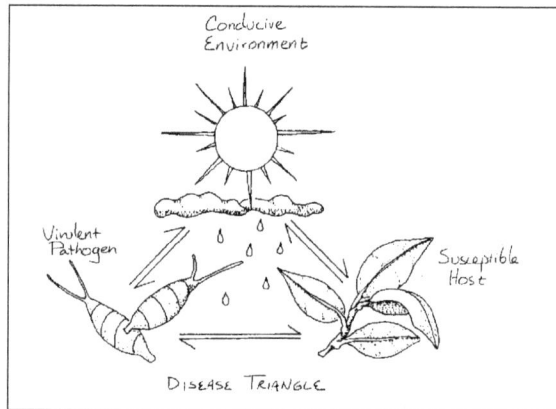

Plant disease development requires a virulent pathogen, a susceptible host, and a conducive environment.

Symptomology

Disease is a negative disruption of the normal functions or structure of a plant, caused by a persistent agent. We know a plant is diseased because of the symptoms it expresses. Symptoms are responses of the host to the pathogen, and may include death, wilting, necrosis (death of tissues), chlorosis (loss of green in tissue), stunting, cankering, wood decay, abnormal growth, fruit loss, and many others. In addition to plant symptoms, signs of the pathogen are often visible. Signs include reproductive structures or mycelium of fungi, or bacterial ooze that can be seen associated with disease symptoms.

Proof of Pathogenicity

Specific combinations of signs and symptoms are often diagnostic for particular kinds of diseases. However, it is important to remember that finding signs of a given fungus associated with diseased tissue does not necessarily mean that that fungus caused the disease. In order to establish that a given set of symptoms is due to a particular micro-organism, it is necessary to complete Koch's Proof of Pathogenicity, where the pathogen is isolated into pure culture and then used to cause disease in a healthy plant. This is the only reliable way of assigning cause of a disease to the appropriate pathogen, but it does have its limitations. Whereas for many diseases isolating the pathogen in pure culture and inoculating hosts is relatively easy, some obligate pathogens (e.g., rusts) cannot be grown in pure culture. For others, reproducing (or even recognizing) the environmental conditions appropriate for disease development may be extremely difficult. Some diseases may take many years for symptoms to develop (particularly true for some tree diseases). In the case of rare or endangered species, infecting hosts experimentally may be unethical. Disease complexes, which have multiple agents acting in concert, are particularly difficult to manipulate experimentally. In each of these cases, Koch's Proof of Pathogenicity is impractical, and a body of evidence in its entirety must be considered in assigning cause to a particular pathogen.

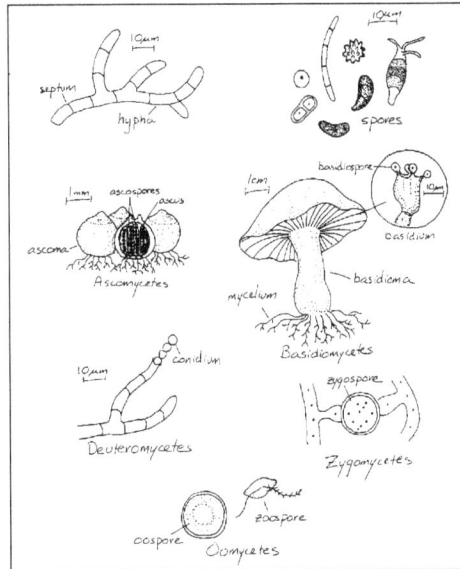

Koch's proof of pathogenicity is essential for showing what causes a disease. Constant association between a pathogen and symptoms, isolation of the pathogen into pure culture, reproduction of symptoms after inoculation of a healthy host, and re-isolation of the pathogen are the key steps.

Disease Cycles

In order to understand where, when, and how a disease will affect tropical forest communities it is helpful to look at the key phases in the disease cycle. Each host and pathogen combination is unique, and it is an error to say "plant diseases work like this" - because there are more exceptions than rules. However, all diseases have several processes in common: dissemination of the pathogen, propagule germination, penetration and infection of the host, the host response, and growth and reproduction of the pathogen, which lead back to dissemination. The key to understanding the ecology of a particular disease lies in the details of how particular pathogens move about and infect their hosts, how different hosts respond, and how the environment influences these processes.

Dissemination

Spores are the primary means of dissemination of pathogens from one host individual to another. Spores may be powdery or have appendages to increase their buoyancy in air or water currents, be flagellated to permit swimming through water, or be borne in sticky masses to facilitate transmission by insect vectors. Large, thick-walled, pigmented spores are likely to be resting structures for survival during adverse environmental conditions, and not be very important in pathogen movement. Looking at the kinds of spores produced and where they are produced (i.e., on the surface of a leaf or embedded in stem tissue) provides clues to where the pathogen is most likely to be successful. For instance, Oomycetes produce flagellated zoospores that require large, water-filled pores in the soil to move from one host to another; thus we would expect that diseases caused by Oomycetes (such as damping-off of seedlings) would be most severe in areas where the soil periodically becomes saturated, and minimal in sunny areas like forest gaps where the soil dries out quickly after a rain. For example, dampingoff of seedlings of Platypodium elegans (Fabaceae) was least severe in canopy-gap environments in a lowland tropical moist forest in Panama.

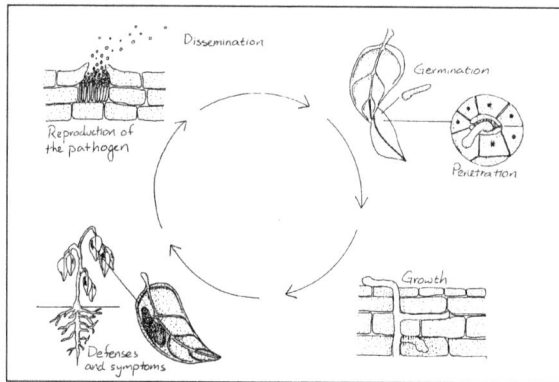

The life cycle of a typical plant disease involves dissemination, germination of spores, penetration and growth through the host, a host response, and reproduction of the pathogen.

Germination

In many cases, however, spore germination and subsequent infection of the host may be far more important than spore dissemination in the epidemiology of plant diseases. In deed, plants in tropical moist forests are inundated by a constant "spore rain". On Barro Colorado Island (BCI) in Panama, Petri plates filled with fungal culture media and placed in the forest understory on a still, dry day in the rainy season caught 9.3 ± 3.3 spores cm-2 hr-1; this includes only readily culturable fungi with air-borne spores, many of which are not plant pathogens (G. S. Gilbert, unpublished data). Extrapolating across a 24-hour period, a 50- cm2 plant leaf would receive approximately 11,000 spores per day! Clearly, only a small fraction of these spores germinates and successfully infects a given leaf. The vast majority blow or wash off of the leaf surface, or die in place from desiccation or ultraviolet radiation. Of those that remain only the subset that encounters the appropriate environmental conditions will germinate (often a particular range of moisture or temperature is critical).

Penetration

Among the spores that do germinate on a leaf surface only a small fraction are likely to be able to breach the first formidable host defense: the waxy leaf cuticle. Pathogens may take three routes to penetrating the cuticle of a host: first, direct penetration, by excreting enzymes or by physical entry via a modified hypha called a "penetration peg"; second, entry through leaf stomata; and third, entry through existing wounds in the leaf. The third route, and probably the most important in tropical forests, is entry through wounds from herbivores or perhaps other pathogens. Garcia-Guzman and Dirzo showed that for both understory and canopy leaves in forest at Los Tuxtlas, Mexico, natural or simulated herbivory was required for nearly all foliar pathogens to cause disease. Similarly, many fungi that attack the woody parts of trees require wounds to gain entry into the host.

Host Response

Once the fungus penetrates the plant a dynamic interplay between the host and symbiont begins. The host may have many defenses, or resistance mechanisms, against infection by microbes. Some of these defenses are constitutive, that is they are a constant feature of the plant's anatomy or physiology. Constitutive defenses include tissue toughness, phenolic compounds and other resins, and nutritional quality of the host, and often are effective against a very broad range of organisms. The other class of resistance mechanisms are induced defenses, which are produced only in response to damage or infection by a particular agent (the elicitor) and include the production of a wide array of toxic secondary compounds. When induced defenses lead to a broad, physiological immunity from further pathogen attack it is known as systemic acquired resistance (SAR). Induction of defenses often requires that the host recognize the invading microorganism; in many cases plants have specific resistance genes which confer the ability to recognize and prevent infection from fungi unless the fungus possesses a particular corresponding gene that allows it to evade the host response, making it virulent on that host. This correspondence between genes for resistance and virulence is called the Gene-for-Gene hypothesis, and is of fundamental importance in understanding the evolution of plantpathogen interactions as well as in breeding for disease-resistant plants. Other forms of disease resistance are more complex, involving multiple loci or with strong interactions between environmental conditions and expression. Such "quantitative disease resistance" is only now yielding to study through molecular methods, but may ultimately prove to be more important than traditional gene-forgene interactions. Differences exist among plant species, among populations, and even among individuals in the resistance mechanisms they possess. Similarly, a particular pathogen may be able to infect a wide range of host species, or be restricted to one or a few closely related hosts. Degree of host specificity is a key issue in determining the impact of a pathogen on a forest community.

Pathogens that evade initial host defenses may cause disease. Disease symptoms are host responses to the pathogen, but may be either host attempts to restrict the pathogen or manipulations of the host by the pathogen. Necrosis of plant parts results in the loss of photosynthetic tissue, usually considered a negative impact on the host. But localized killing of host tissue (called a hypersensitive response) can actually limit the spread of a biotrophic pathogen through the host by cutting the pathogen off from necessary nutrients: thus necrosis may in some cases be a type of defense. In other cases some pathogenic fungi that are unable to efficiently colonize living tissue will instead secrete toxins that kill plant cells in advance of the growing hyphae, allowing the fungus to invade poorly defended necrotic tissue. Pathogens may spread through the host by colonizing living tissue, killing tissue and growing into it, or by using the host vascular system as conduits for rapid movement.

Not all fungi that penetrate a leaf or twig will cause disease. Nearly every healthy-appearing leaf in a tropical forest is infected with a diversity of fungi, although the number of infections per leaf may vary widely. Fungi that live inside a plant without causing disease symptoms are called endophytes. These endophytes may be participants in various kinds of symbioses. Some fungi are known to provide benefits to the host plant through protection from herbivores or diseases. Others may be commensals that remain quiescent within the host until it dies, and then consume the dead material as a saprotroph. Many pathogens may pass a long, latent endophytic phase before causing disease in response to host stress or senescence. For instance, the Ascomycete Botryosphaeria dothidea may pass many months in a latent phase after infecting stems of Tetragastris panamensis (Burseraceae); however, cankers develop rapidly on infected individuals soon after the onset of drought. Other factors such as nutrient deficiency, air pollution, attack from insects, or light stress could similarly induce disease development.

Reproduction

In all cases the microbial symbiont must eventually reproduce in order to colonize a new host individual. Reproduction for fungi involves first developing sufficient mycelium and energy reserves to support reproduction, and then producing spores. Spores may be produced and liberated either from a living host providing inoculum for infecting other parts of the same plant or other individuals, or may be produced only after the death of the host. How quickly a pathogen reproduces and how much inoculum it produces are key determinants of the rate a disease spreads through a host population and evolves responses to host defenses. The life cycle of many microbes is much faster than the life cycle of the plants that they infect; in some cases a pathogen can pass through the cycle from infection to reproduction in only days to weeks, or even pass through several generations on a single host individual. This provides much greater opportunity for the pathogen to evolve mechanisms to overcome host resistance than for the host to overcome new virulence factors.

Risks, Conservation and Forest Management

Microorganisms thus play critical roles in the dynamics, diversity, and processes in tropical forests. Mutualistic symbioses between bacteria or fungi and plants can provide competitive advantages to the hosts and alter community composition. Endemic diseases may play crucial roles in maintaining community species diversity and provide habitat heterogeneity for wildlife. However, human intervention can often lead to the development of devastating disease epidemics. The introduction of pathogens from other ecosystems can have dramatic impacts on native vegetation. Forest disturbance, through logging or other human activities, can stress plants and create wounds leading to large increases in disease levels. Forest fragmentation can change microclimatic conditions in the forests, leading to stresses that can incite disease development. Forestry practices such as plantations that result in high-density stands of very low diversity are especially susceptible to disease epidemics, through density-dependent disease development. Plantation forests and monocultural agriculture may pose threats to nearby natural forests by permitting the increase in pathogen inoculum that, given susceptible hosts in the native vegetation, may infect the natural stands from an inoculum reservoir.

Plant diseases must be taken into consideration in the planning of forest reserves. Because diseases are contagious and the spread from individual to individual is dependent on the proximity between individuals, density, diversity, and connectivity are all important elements in determining if particular forest areas are vulnerable to catastrophic epidemics. Small forest reserves are subject to dramatic changes should a pathogen be introduced or conditions change to promote disease development. Much of the remaining tropical forests are in discontinuous fragments, and there are efforts to provide biological corridors to connect them, permitting gene flow among populations and reducing the dangers from stochastic effects. Although these are important benefits, the corridors also increase the risk that a disease that develops in one fragment can spread along corridors to other forest fragments. Such risks have been discussed for animal diseases, but to date there are no data on the risks from plant diseases that corridors carry.

The diversity of plant-microorganism interactions in tropical forests may provide economic incentives for forest conservation, as well. Biological prospecting for fungi with industrial uses in biocontrol, cosmetics, and pharmaceuticals may be even more promising than plant-centered bioprospecting. Fungi have traditionally been sources of particularly creative secondary chemicals, and because many fungi can be grown in culture, they are well-suited for industrial exploitation. Fungi collected in initial sampling expeditions can be maintained in collections and grown quickly and inexpensively through fermentation technology, eliminating the need for repeated expeditions to collect additional materials. Fungal collections can be especially rich source of information because many more samples of fungi than plant materials can be collected and processed with the same effort, easily providing thousands of different fungal types for automated industrial screening. Industrial interest in access to diverse tropical ecosystems for microbial bioprospecting may be one tool for leveraging resources for forest conservation. Merck, Inc. in Costa Rica and Novartis, Inc.

in Panama, Mexico, and India are among the current partnerships between pharmaceutical companies and tropical nations using fungal bioprospecting to help promote forest conservation.

Sustainable Agriculture and Plant Diseases

Agriculture is changing fast and with it the landscape through which disease spreads. This imposes new demands on our understanding of epidemiology if we are to control disease efficiently, whether by genetical, chemical, biological or cultural means. The sorts of questions that need to be addressed are focused on discovering the factors that influence the invasion and persistence of new pathogenic strains, how and why they outcompete resident pathogens and how to promote durable methods of control. This requires an understanding of what controls the variability of epidemics between one location and another and from one season to another, and how this impinges upon local, national and sometimes international crop loss.

Why is agriculture changing so fast? What is the evidence? In large parts of the world, intensively managed farms are becoming larger, interspersed with smaller, organically and conventionally managed farms with diverse livestock and cropping patterns. Global warming is changing the national and international ranges of pests and disease. Economic pressures and global trade are changing national cropping patterns. It is anticipated that the demand for cereals will increase by 20% by 2020 as world population grows. The corresponding increase in demand for animal products is estimated to be 50%, in response to increasing affluence and urbanization, notably in southeast Asia. The recent accession of 10 new states into the EU along with reform of the Common Agricultural Policy is likely to change cropping patterns in Western Europe. Novel crops for biofuel, plastics and intensive specialized production of pharmaceutical crops under glass are probable. Meanwhile, our understanding of the genetical and chemical bases of disease control is accelerating following investment in molecular biology. The costs, though, for release of new varieties and for the development and registration of new chemicals have escalated. The quest for durable control itself rests on a paradox. Since most plants are self-evidently resistant to most pathogens, it seems perfectly reasonable to assume that advancing knowledge of the molecular and cellular bases of host–pathogen interaction will identify the means not only to engineer or to select durable resistance but also to produce effective and environmentally neutral forms of chemical control. Yet failures still occur, whether from the release of novel resistance genes or from new pesticides and fungicides.

Agriculture continues to be confronted with new and recurrent epidemics. Notable examples include recent epidemics of cassava mosaic virus in West Africa, citrus canker in Florida, rhizomania disease of sugar beet in the UK and Western Europe, and the arrival and spread of Asian soya bean rust, caused by Phakopsora pachyrhizi, in South

America and subsequent entry into the United States. Similar problems arise in natural and semi-natural communities, where the effects are especially noticeable on forest and amenity trees: sudden oak death, caused by Phytophthora ramorum is spreading rapidly on coast live oak and tanoak in California, for which the probable strategy for control is sanitation and containment. Meanwhile, Dutch elm disease may recur in the UK as may chestnut blight caused by Cryphonectria parasitica in the US, for each of which there is continued interest in biological control by RNA viruses.

Most forms of disease control are screened for effectiveness at the small scale. Often this is done at scales as small as the single plant for initial screening, though more usually it involves field plots and ultimately fields. Yet successful deployment, and the risk of failure, occurs at scales much larger than this, at the regional, national or even international scales. We can reconcile these scales using an epidemiological framework that allows us to predict how measurable changes in latent and infectious periods or transmission rate might affect the regional spread of disease. This, in turn, requires fusion of population genetics and epidemiology at scales extending from the field to the landscape and even to continental deployment of control measures. Progress in sustainable disease control measures also requires an understanding of economic and social constraints. These are all too frequently ignored in epidemiological models, while economic models are often biologically naive, failing particularly to allow for the dynamical nature of most epidemics.

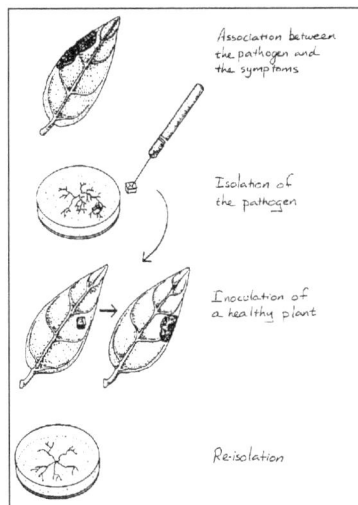

Schematic relationships of selected components of an epidemiological framework encompassing population dynamic, population genetic, landscape dynamics and economic approaches.

An epidemiological framework for sustainable disease control requires a suite of models to analyse and predict the effects of control on the spatial and temporal dynamics of disease, together with methods to parametrize the disease. One important switch of emphasis is from within-field to regional control of disease. Underlying this analysis is a shift in emphasis from private to public benefit, whereby optimization of control strategies may increasingly be exercised from a regional perspective over a population of growers rather than seeking to optimize control in each field. Many of the components for an epidemiological framework have been individually studied but not always at the same scale and often with little overlap. These include models from both medical and botanical epidemiology (to analyse and predict the effects of control strategies on the spatial and temporal dynamics of disease), population genetics (for the evolution of virulence and pesticide resistance), landscape ecology (for the spatial structure of susceptible host populations) and environmental economics (to allow for cost constraints and decision making under uncertainty). Progress in developing an epidemiological framework for sustainable control of disease also depends upon theoretical advances in mathematics and statistical physics (particularly for transient behaviour, spatially extended dynamics, percolation and network theory) as well as in probability and statistics (for stochastic dynamics and parameter estimation of nonlinear models using modern computer-intensive methods to explore large regions of parameter space). The relationships among some of these are shown schematically in figure. Comprehensive treatments of model structure and analysis are also available in Gilligan for epidemics of fungal and fungal-like diseases and by Madden et al. for virus diseases.

Crop Mosaics, Heterogeneity and Topology of Crops in the Landscape

Much thought has been given to the construction and deployment of crop multilines and mixtures to slow the spread of disease within fields. Following early work by Browning & Frey on multilines and Wolfe on mixtures, interest in the strategies waned, partly due to resistance from processors. Recently, however, interest has been revived in China and Western Europe. Slowing the increase of disease within fields may have a knock-on effect in slowing the spread of disease between fields and, hence, restricting invasion through the landscape. By thinking of fields as subpopulations in a structured metapopulation, the effect of mixtures in the landscape can conveniently be thought of as delaying the transit time between fields. We define the transit time as the time from when the first individual becomes infected in one field and starts an epidemic in another field. Most theoretical work on transit times in metapopulations has focused on persistence. Thus, Swinton showed that above a certain critical subpopulation size, the expected extinction time for a metapopulation scales with the number and size of subpopulations. There is a phase transition (i.e. a switch in behaviour from short to long extinction times) around the critical subpopulation size, NC. Below NC, the time to extinction is very short because the amount of susceptible hosts is not sufficient to maintain the epidemic before it spreads to the next subpopulation. As the population of susceptibles increases above NC, there is a sudden transition to long extinction times. Increasing the numbers of subpopulations in the metapopulation delays extinction but does not affect the critical value of NC at which the phase transition occurs. This has been demonstrated for animal disease and proposed for botanical epidemics, with additional theoretical work on invasion as well as persistence. It is now ready to be tested in the field.

How far and how fast a pathogen spreads through a landscape and, indeed, whether or not it persists and for how long, depends upon the crop mosaic within the landscape. Whether or not invasion occurs depends upon the relative magnitude of the modal dispersal distances and the scale of heterogeneity in the landscape. Considering disease spread in this way is still at an early stage. A convenient starting point is to assume that disease spreads on a lattice and then to introduce gaps and so to advance to spread through a realistic population of fields. For a given pathogen, each field of a susceptible crop is classified as susceptible (i.e. healthy), infected or, if appropriate, recovered. Non-host crops are unavailable to the pathogen and are treated as gaps. The application of a protectant fungicide temporarily renders a susceptible field unavailable for infection. Whether or not a pathogen invades now depends upon the spatial and temporal dynamics of the crop mosaic. A simple analysis based upon ideas from percolation, derived in statistical physics, illustrates an important relationship between dispersal kernels and invasion. When spread occurs between adjacent fields, i.e. by nearest-neighbour spread, the dispersal kernel then describes the probability distribution for transmission of infection between an infected field and its susceptible neighbour. It follows from percolation theory that there is a critical probability above which disease spreads and below which it dies out. By increasing the distance between susceptible fields, it is possible to bring the percolation probability for a given lattice below the threshold and so prevent invasion. These results are stochastic: it follows that there will be some spread when the percolation probability is below the threshold and occasionally invasion may occur but, on average, we would expect the strategy to work, albeit under these idealized conditions.

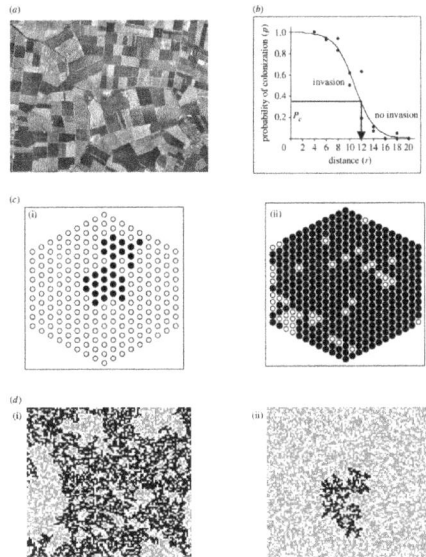

Spread through a landscape with nearest neighbour transmission between infected and susceptible sites.

Suppose now that the density of susceptible crops per unit area of land exceeds the threshold and invasion is expected. Introducing gaps into the lattice by the application of localized chemical control could switch an invasive into a non-invasive mosaic. The fineness of the control and the precision of the phase transition is sufficiently striking to lead us to question whether or not it is necessary to treat all fields in the landscape in order to prevent invasion. The answers from these theoretical investigations would suggest not. The ideas have yet to be tested in the field, however; something that presents significant challenges in large-scale testing of new approaches towards sustainable control. Meanwhile in order to gain more insight into these approaches and, in particular, to understand variability between replicate epidemics, we have tested some of the ideas in a series of microcosm experiments involving R. solani spreading between nutrient sites at different densities with and without gaps.

In figure, (a) Aerial photograph showing a typical example of a UK crop mosaic. (b) Probability distribution for transmission of infection to susceptible sites at different distances from the infected source showing the percolation threshold (P_c) for a triangular lattice. (c) Effects of a small change in the distance between susceptible (open circle) sites (i) above and (ii) below the threshold distance on the invasion of an epidemic measured by the spread of infected (filled circle) sites. (d) Effects of introducing gaps to simulate chemical control in the landscape upon a percolating system, showing the fineness of the control about a phase transition; white squares indicate control, light grey indicate susceptible and dark infected. (b,c) Derived from experimental microcosms. (d) Derived from a computer simulation by kind permission of Dr J. Ludlam.

As an epidemic spreads through a heterogeneous landscape, the dynamical contact structure, which measures the exposure of susceptible to infected sites, also changes. This too can slow down or speed up an epidemic and is a property of the topology of the system. So far, we have discussed susceptibles and gaps. What if there are differential susceptibilities for susceptible crops in a mosaic? This could reflect incomplete chemical control in treated fields or partial genetical resistance. By extending the model experimental system to include spread through mixed populations of radish and mustard in replicated microcosms, we simulated experimentally the spread through a heterogeneous landscape. Not surprisingly, the inclusion of a less susceptible host can slow the rate of spread of an epidemic compared with spread though a homogeneous mosaic. However, the rate of spread changes nonlinearly with the relative densities. If this holds for the large-scale transmission between fields, it follows that the introduction of a critical proportion of a less susceptible crop could slow the spread of infection through the landscape. Moreover, the differential rates reflect not only differences in susceptibility but also differences in infectivity. This means that for two types of host, four types of transmission rates occur, depending upon which host is the donor (infected) or recipient (susceptible). Otten et al. showed how to calculate transmission rates from empirical data which, in turn, can be used to predict the effects of changes in the topology and composition of the mixture. By treating fields as units this approach can, in principle, be extended to gain insight into the effects of different levels and densities of partial control at the landscape scale.

Invasion and Persistence of Virulent and Fungicide-Resistant Strains

Epidemiological models for the invasion and persistence of chemical control are similar to those for genetical control. Until quite recently, most models ignored stochasticity and density dependence imposed by the dynamical changes in the availability of untreated tissue. Moreover, by invoking exponential growth of the pathogen, invasion of fungicide-resistant forms is inevitable, and attention focuses not on whether or not a resistant strain can invade but the time to reach a critical level. This is unrealistic. By allowing for competition for susceptible (untreated) sites, it is possible to show that there is a threshold below which resistance cannot develop within the host population. The threshold depends upon the relative fitness of the resistant and sensitive strains and the effectiveness of control. The latter may influence one or more of the following: reduction in the transmission of infection, the duration of infectiousness and the conversion of host to infectious pathogen tissue. The relative fitness of the resistant strain is given by the ratio R_{or}/R_{os}, where Ros and Ror are the basic reproductive numbers for the sensitive and resistant strains, respectively. Even crude estimates for the efficiency of control and the basic reproductive numbers can provide a simple rule of thumb for whether or not invasion may be expected.

Fungicides differ from many forms of genetical control in being subject to decay over time, requiring repeated application. By including parameters for the amount of fungicide applied, longevity and application frequency of the chemical, it is possible to predict the outcome of invasion of the resistant strain, even when there is a time-varying selection pressure on the pathogen population. Notably, these models share a common generic structure with antiviral drug resistance and antibiotic resistance.

The durability of genetical and chemical control depends not only on the ability of a virulent or fungicide-sensitive strain to invade an existing pathogen population but also whether the wild type is completely replaced or can coexist. Complete exclusion of the resident (controllable) pathogen strain by the invading strain means that the gene or fungicide cannot be used again as a sole method of control. Empirical evidence shows both scenarios with coexistence and competitive exclusion. To develop effective resistance management strategies, it is imperative to understand the processes that influence which of these outcomes is likely to occur. Initially, it was thought that models could be of little assistance because coexistence was predicted to occur only under exceptional circumstances with restrictive assumptions about the magnitudes of the parameter values. The obvious way to account for a spectrum from competitive exclusion to coexistence is to allow for spatial heterogeneity in the selection pressures. This is conveniently done using a metapopulation framework. Thus, when Parnell et al. allowed for differential selection pressures due to incomplete coverage of plants by fungicide either within or between crops, three scenarios were possible, ranging from failure to invade, through coexistence, to competitive exclusion of the resident strain. The outcome within fields depends upon the balance between incomplete spray coverage and a cost to resistance, with healthy host density maximized when there is coexistence. Recent work by Salathé et al. suggests that coexistence is possible within stochastic systems without invoking costs of resistance.

At the regional scale, invasion of resistant strains is determined by a trade-off between the fraction of fields that are sprayed and the intrinsic reproductive ability Ro of the target pathogen. If the between-field movement of the pathogen is high (high Ro), the resistant strain dominates all treated fields but the sensitive strain dominates all untreated fields. This occurs because, in the long term, resistant strains are competitively superior in treated fields and sensitive strains are competitively superior in untreated fields. If Ro is very high, mixing is complete and all treated fields become infested with the resistant strain and all untreated fields become infested with the sensitive strain. If, however, Ro is low, strains cannot move between fields to the extent that allows them to capitalize on their within-field competitive advantage and may therefore be excluded. The outcome is then dependent on the fraction of fields sprayed. Once again the analogy with refugia is apparent.

Further work on the invasion and persistence of virulent and fungicide-resistant pathogen strains will increasingly focus on the integration of genetical and epidemiological mechanisms. Foremost among these will be empirical evidence or otherwise for fitness costs as well as research on the evolution of fitness modifiers and evolutionary trade-offs among epidemiological parameters. Clearly, there is also a need to identify control strategies that balance the conflicting aims of resistance management (to reduce the risk of failure) and yield enhancement by application of genetical, chemical and other control methods to suppress disease. Some initial models that integrate crop yield have been explored by Hall et al. Most strategies depend upon host diversification within the landscape. We do not, however, know how introducing host diversity into the landscape affects evolutionary divergence towards specialist or generalist pathogens, or even if it might lead to a switch in pathogenicity from one host to another. This can be investigated by analysing the evolutionary trade-offs that occur over successive generations of a pathogen exposed to two or more hosts. Preliminary results show that evolutionary outcomes strongly depend on the shape of the trade-off curve between pathogen transmissions on sympatric hosts. Using methods based upon adaptive dynamics, it has been possible to determine criteria under which evolutionary branching occurs from a monomorphic into a dimorphic population, as well as the conditions that lead to the evolution of specialist (single host range) or generalist (multiple host range) pathogen populations. Since some pathogen species can undergo 20–30 generations in a growing season, the consequences of this form of evolution may become apparent within decades.

Optimization of Disease Control using an Epidemiological Framework

The control of disease epidemics often requires expensive resources. These include investment costs for breeding programmes, the development, testing and registration of new chemicals or biological control agents. There are also variable costs for the application of control methods. There may be economic, environmental or ecological restrictions on the use of different methods associated with the accumulation of toxic chemicals in the environment, or the risk of failure through premature build-up of virulence to resistance genes or insensitivity to pesticides in pathogen populations that could render the control and investment ineffective. This creates two problems for the implementation of control. The first is a strategic issue about the long-term effectiveness and the corresponding risks of failure associated with different control strategies. We discuss briefly how the risks can be ameliorated by buffered implementation of novel control. The second is how to optimize the deployment of control when resources are limited or there are other restrictions on use (such as the risk of failure through over-use).

Buffered Implementation of Genetic and Chemical Control

Following some spectacular failures, increasing attention is being given to the buffered implementation of genetical and chemical control methods through time and space. This is variously done by the cultivation of refugia of non-resistant or untreated crops, by alternating pesticides through time and space to impose different selection pressures, and by diversifying the genetic bases for control. Each demands an understanding of the 'dynamical landscape' in order to match the scale of control with the scale of the epidemic. This is most well advanced for a pest problem involving the deployment of Bt-resistance for insect pests on maize and cotton in the US and elsewhere, where there are mandatory requirements for refugia. By permitting multiplication and persistence of the sensitive form of the pest population on susceptible crops within refugia, the build-up of mutant pests that can feed and reproduce on the Bt-resistant crops is delayed. Arguably though, with this and other schemes most attention has focused on relatively local scales, yet there may be broader issues for larger scales. For example, should a new pesticide or resistant variety be released at a continental scale? Would heterogeneity at a local scale be sufficient to prevent problems of invasion of pesticide-resistant or virulent strains at a much larger scale? We do not yet know. Recent examples for the apparent widescale occurrence and spread throughout Western Europe of fungicide resistance forms in the eyespot and mildew pathogens of wheat suggests that large-scale dynamics must be considered. These pathogens differ in dispersal mode, with eyespot predominantly spread by short-distance rainsplash whereas mildew can spread over longer distances by wind. A combination of the analyses of Aylor for continental spread with those of network and landscape models may help to answer these questions.

Biological Control of Plant Pathogens

The terms "biological control" and its abbreviated synonym "biocontrol" have been used in different fields of biology, most notably entomology and plant pathology. In entomology, it has been used to describe the use of live predatory insects, entomopathogenic nematodes, or microbial pathogens to suppress populations of different pest insects. In plant pathology, the term applies to the use of microbial antagonists to suppress diseases as well as the use of host-specific pathogens to control weed populations. In both fields, the organism that suppresses the pest or pathogen is referred to as the biological control agent (BCA). More broadly, the term biological control also has been applied to the use of the natural products extracted or fermented from various sources. These formulations may be very simple mixtures of natural ingredients with specific activities or complex mixtures with multiple effects on the host as well as the target pest or pathogen. And, while such inputs may mimic the activities of living organisms, non-living inputs should more properly be referred to as biopesticides or biofertilizers, depending on the primary benefit provided to the host plant. The various definitions offered in the scientific literature have sometimes caused confusion and controversy. For example, members of the U.S. National Research Council took into account modern biotechnological developments and referred to biological control as "the use of natural or modified organisms, genes, or gene products, to reduce the effects of undesirable organisms and to favor desirable organisms such as crops, beneficial insects, and microorganisms", but this definition spurred much subsequent debate and it was frequently considered too broad by many scientists who worked in the field. Because the term biological control can refer to a spectrum of ideas, it is important to stipulate the breadth of the term when it is applied to the review of any particular work.

Published definitions of biocontrol differ depending on the target of suppression; number, type and source of biological agents; and the degree and timing of human intervention. Most broadly, biological control is the suppression of damaging activities of one organism by one or more other organisms, often referred to as natural enemies. With regards to plant diseases, suppression can be accomplished in many ways. If growers' activities are considered relevant, cultural practices such as the use of rotations and planting of disease resistant cultivars (whether naturally selected or genetically engineered) would be included in the definition. Because the plant host responds to numerous biological factors, pathogenic and non-pathogenic, induced host resistance might be considered a form of biological control. More narrowly, biological control refers to the purposeful utilization of introduced or resident living organisms, other than disease resistant host plants, to suppress the activities and populations of one or more plant pathogens. This may involve the use of microbial inoculants to suppress a single type or class of plant diseases. Or, this may involve managing soils to promote the combined activities of native soil- and plant-associated organisms that contribute to general suppression. Most narrowly, biological control refers to the suppression of a single pathogen (or pest), by a single antagonist, in a single cropping system. Most specialists in the field would concur with one of the narrower definitions presented above.

Types of Interactions Contributing to Biological Control

Throughout their lifecycle, plants and pathogens interact with a wide variety of organisms. These interactions can significantly affect plant health in various ways. In order to understand the mechanisms of biological control, it is helpful to appreciate the different ways that organisms interact. Note, too, that in order to interact, organisms must have some form of direct or indirect contact. Odum proposed that the interactions of two populations be defined by the outcomes for each. The types of interactions were referred to as mutualism, protocooperation, commensalism, neutralism, competition, amensalism, parasitism, and predation. While the terminology was developed for macroecology, examples of all of these types of interactions can be found in the natural world at both the macroscopic and microscopic level. And, because the development of plant diseases involves both plants and microbes, the interactions that lead to biological control take place at multiple levels of scale.

From the plant's perspective, biological control can be considered a net positive result arising from a variety of specific and non-specific interactions. Using the spectrum of Odum's concepts, we can begin to classify and functionally delineate the diverse components of ecosystems that contribute to biocontrol. Mutualism is an association between two or more species where both species derive benefit. Sometimes, it is an obligatory lifelong interaction involving close physical and biochemical contact, such as those between plants and mycorrhizal fungi. However, they are generally facultative and opportunistic. For example, bacteria in the genus Rhizobium can reproduce either in the soil or, to a much greater degree, through their mutualistic association with legume plants. These types of mutualism can contribute to biological control, by fortifying the plant with improved nutrition and/or by stimulating host defenses. Protocooperation is a form of mutualism, but the organisms involved do not depend exclusively on each other for survival. Many of the microbes isolated and classified as BCAs can be considered facultative mutualists involved in protocooperation, because survival rarely depends on any specific host and disease suppression will vary depending on the prevailing environmental conditions. Further down the spectrum, commensalism is a symbiotic interaction between two living organisms, where one organism benefits and the other is neither harmed nor benefited. Most plant-associated microbes are assumed to be commensals with regards to the host plant, because their presence, individually or in total, rarely results in overtly positive or negative consequences to the plant. And, while their presence may present a variety of challenges to an infecting pathogen, an absence of measurable decrease in pathogen infection or disease severity is indicative of commensal interactions. Neutralism describes the biological interactions when the population density of one species has absolutely no effect whatsoever on the other. Related to biological control, an inability to associate the population dynamics of pathogen with that of another organism would indicate neutralism.

In contrast, antagonism between organisms results in a negative outcome for one or both. Competition within and between species results in decreased growth, activity and/or fecundity of the interacting organisms. Biocontrol can occur when non-pathogens compete with pathogens for nutrients in and around the host plant. Direct interactions that benefit one population at the expense of another also affect our understanding of biological control. Parasitism is a symbiosis in which two phylogenetically unrelated organisms coexist over a prolonged period of time. In this type of association, one organism, usually the physically smaller of the two (called the parasite) benefits and the other (called the host) is harmed to some measurable extent. The activities of various hyperparasites, i.e., those agents that parasitize plant pathogens, can result in biocontrol. And, interestingly, host infection and parasitism by relatively avirulent pathogens may lead to biocontrol of more virulent pathogens through the stimulation of host defense systems. Lastly, predation refers to the hunting and killing of one organism by another for consumption and sustenance. While the term predator typically refer to animals that feed at higher trophic levels in the macroscopic world, it has also been applied to the actions of microbes, e.g. protists, and mesofauna, e.g. fungal feeding nematodes and microarthropods, that consume pathogen biomass for sustenance. Biological control can result in varying degrees from all of these types of interactions, depending on the environmental context within which they occur. Significant biological control, as defined above, most generally arises from manipulating mutualisms between microbes and their plant hosts or from manipulating antagonisms between microbes and pathogens.

Mechanisms of Biological Control

Because biological control can result from many different types of interactions between organisms, researchers have focused on characterizing the mechanisms operating in different experimental situations. In all cases, pathogens are antagonized by the presence and activities of other organisms that they encounter. Here, we assert that the different mechanisms of antagonism occur across a spectrum of directionality related to the amount of interspecies contact and specificity of the interactions. Direct antagonism results from physical contact and/or a high-degree of selectivity for the pathogen by the mechanism(s) expressed by the BCA(s). In such a scheme, hyperparasitism by obligate parasites of a plant pathogen would be considered the most direct type of antagonism because the activities of no other organism would be required to exert a suppressive effect. In contrast, indirect antagonisms result from activities that do not involve sensing or targeting a pathogen by the BCA(s). Stimulation of plant host defense pathways by non-pathogenic BCAs is the most indirect form of antagonism. However, in the context of the natural environment, most described mechanisms of pathogen suppression will be modulated by the relative occurrence of other organisms in addition to the pathogen. While many investigations have attempted to establish the importance of specific mechanisms of biocontrol to particular pathosystems, all of the mechanisms described below are likely to be operating to some extent in all natural and managed ecosystems. And, the most effective BCAs studied to date appear to antagonize pathogens using multiple mechanisms. For instance, pseudomonads known to produce the antibiotic 2,4-diacetylphloroglucinol (DAPG) may also induce host defenses. Additionally, DAPG-producers can aggressively colonize roots, a trait that might further contribute to their ability to suppress pathogen activity in the rhizosphere of wheat through competition for organic nutrients.

Table: Types of interspecies antagonisms leading to biological control of plant pathogens.

Type	Mechanism	Examples
Direct antagonism	Hyperparasitism/predation	Lytic/some nonlytic mycoviruses Ampelomyces quisqualis Lysobacter enzymogenes Pasteuria penetrans Trichoderma virens
Mixed-path antagonism	Antibiotics	2,4-diacetylphloroglucinol Phenazines Cyclic lipopeptides
	Lytic enzymes	Chitinases Glucanases Proteases
	Unregulated waste products	Ammonia Carbon dioxide Hydrogen cyanide
	Physical/chemical interference	Blockage of soil pores Germination signals consumption Molecular cross-talk confused
Indirect antagonism	Competition	Exudates/leachates consumption Siderophore scavenging Physical niche occupation
	Induction of host resistance	Contact with fungal cell walls Detection of pathogen-associated, molecular patterns Phytohormone-mediated induction

Hyperparasites and Predation

In hyperparasitism, the pathogen is directly attacked by a specific BCA that kills it or its propagules. In general, there are four major classes of hyperparasites: obligate bacterial pathogens, hypoviruses, facultative parasites, and predators. Pasteuria penetrans is an obligate bacterial pathogen of root-knot nematodes that has been used as a BCA. Hypoviruses are hyperparasites. A classic example is the virus that infects Cryphonectria parasitica, a fungus causing chestnut blight, which causes hypovirulence, a reduction in disease-producing capacity of the pathogen. The phenomenon has controlled the chestnut blight in many places. However, the interaction of virus, fungus, tree, and environment determines the success or failure of hypovirulence. There are several fungal parasites of plant pathogens, including those that attack sclerotia (e.g. Coniothyrium minitans) while others attack living hyphae (e.g. Pythium oligandrum). And, a single fungal pathogen can be attacked by multiple hyperparasites. For example, Acremonium alternatum, Acrodontium crateriforme, Ampelomyces quisqualis,

Cladosporium oxysporum, and Gliocladium virens are just a few of the fungi that have the capacity to parasitize powdery mildew pathogens. Other hyperparasites attack plant-pathogenic nematodes during different stages of their life cycles (e.g. Paecilomyces lilacinus and Dactylella oviparasitica). In contrast to hyperparasitism, microbial predation is more general and pathogen non-specific and generally provides less predictable levels of disease control. Some BCAs exhibit predatory behavior under nutrient-limited conditions. However,Trichoderma produce a range of enzymes that are directed against cell walls of fungi. However, when fresh bark is used in composts, Trichoderma spp. do not directly attack the plant pathogen, Rhizoctonia solani. But in decomposing bark, the concentration of readily available cellulose decreases and this activates the chitinase genes of Trichoderma spp., which in turn produce chitinase to parasitize R. solani.

Antibiotic-mediated Suppression

Antibiotics are microbial toxins that can, at low concentrations, poison or kill other microorganisms. Most microbes produce and secrete one or more compounds with antibiotic activity. In some instances, antibiotics produced by microorganisms have been shown to be particularly effective at suppressing plant pathogens and the diseases they cause. Some examples of antibiotics reported to be involved in plant pathogen suppression are listed in table. In all cases, the antibiotics have been shown to be particularly effective at suppressing growth of the target pathogen in vitro and/or in situ. To be effective, antibiotics must be produced in sufficient quantities near the pathogen to result in a biocontrol effect. In situ production of antibiotics by several different biocontrol agents has been measured; however, the effective quantities are difficult to estimate because of the small quantities produced relative to the other, less toxic, organic compounds present in the phytosphere. And while methods have been developed to ascertain when and where biocontrol agents may produce antibiotics, detecting expression in the infection court is difficult because of the heterogenous distribution of plant-associated microbes and the potential sites of infection. In a few cases, the relative importance of antibiotic production by biocontrol bacteria has been demonstrated, where one or more genes responsible for biosynthesis of the antibiotics have been manipulated.

For example, mutant strains incapable of producing phenazines or phloroglucinols have been shown to be equally capable of colonizing the rhizosphere but much less capable of suppressing soilborne root diseases than the corresponding wild-type and complemented mutant strains. Several biocontrol strains are known to produce multiple antibiotics which can suppress one or more pathogens. For example, Bacillus cereus strain UW85 is known to produce both zwittermycin and kanosamine. The ability to produce multiple antibiotics probably helps to suppress diverse microbial competitors, some of which are likely to be plant pathogens. The ability to produce multiple classes of antibiotics, that differentially inhibit different pathogens, is likely to enhance biological control. More recently, Pseudomonas putida WCS358r strains genetically engineered to produce phenazine and DAPG displayed improved capacities to suppress plant diseases in field-grown wheat.

Table: Some of antibiotics produced by BCAs.

Antibiotic	Source	Target pathogen	Disease
2, 4-diacetyl-phlo-roglucinol	Pseudomonas fluorescens F113	Pythium spp.	Damping off
Agrocin 84	Agrobacterium radiobacter	Agrobacterium tumefaciens	Crown gall
Bacillomycin D	Bacillus subtilisAU195	Aspergillus flavus	Aflatoxin contamination
Bacillomycin, fengycin	Bacillus amyloliquefaciensFZB42	Fusarium oxysporum	Wilt
Xanthobaccin A	Lysobacter sp. strain SB-K88	Aphanomyces cochlioides	
Gliotoxin	Trichoderma virens	Rhizoctonia solani	Root rots
Herbicolin	Pantoea agglomeransC9-1	Erwinia amylovora	Fire blight
Iturin A	B. subtilis QST713	Botrytis cinerea and R. solani	Damping off
Mycosubtilin	B. subtilis BBG100	Pythium aphanidermatum	Damping off
Phenazines	P. fluorescens 2-79 and 30-84	Gaeumannomyces graminis var. tritici	Take-all
Pyoluteorin, pyrrolnitrin	P. fluorescens Pf-5	Pythium ultimum and R. solani	Damping off
Pyrrolnitrin, pseudane	Burkholderia cepacia	R. solani and Pyricularia oryzae	Damping off and rice blast
Zwittermicin A	Bacillus cereus UW85	Phytophthora medicaginis and P. aphanidermatum	Damping off

Lytic Enzymes and other Byproducts of Microbial Life

Diverse microorganisms secrete and excrete other metabolites that can interfere with pathogen growth and/or activities. Many microorganisms produce and release lytic enzymes that can hydrolyze a wide variety of polymeric compounds, including chitin, proteins, cellulose, hemicellulose, and DNA. Expression and secretion of these enzymes by different microbes can sometimes result in the suppression of plant pathogen activities directly. For example, control of Sclerotium rolfsii by Serratia marcescens appeared to be mediated by chitinase expression. And, a b-1,3-glucanase contributes significantly to biocontrol activities of Lysobacter enzymogenes strain C3. While they may stress and/or lyse cell walls of living organisms, these enzymes generally act to decompose plant residues and nonliving organic matter. Currently, it is unclear how much of the lytic enzyme activity that can be detected in the natural environment represents specific responses to microbe-microbe interactions. It seems more likely that such activities are largely indicative of the need to degrade complex polymers in order to obtain carbon nutrition. Nonetheless, microbes that show a preference for colonizing and lysing plant pathogens might be classified as biocontrol agents. Lysobacter and Myxobacteria are known to produce copious amounts of lytic enzymes, and some isolates have been shown to be effective at suppressing fungal plant pathogens. So, the lines between competition, hyperparasitism, and antibiosis are generally blurred. Furthermore, some products of lytic enzyme activity may contribute to indirect disease suppression. For example, oligosaccharides derived from fungal cell walls are known to be potent inducers of plant host defenses. Interestingly, Lysobacter enzymogenes strain C3 has been shown to induce plant host resistance to disease, though the precise activities leading to this induction are not entirely clear. The quantitative contribution of any and all of the above compounds to disease suppression is likely to be dependent on the composition and carbon to nitrogen ratio of the soil organic matter that serves as a food source for microbial populations in the soil and rhizosphere. However, such activities can be manipulated so as to result in greater disease suppression. For example, in post-harvest disease control, addition of chitosan can stimulate microbial degradation of pathogens similar to that of an applied hyperparasite. Chitosan is a non-toxic and biodegradable polymer of beta-1,4-glucosamine produced from chitin by alkaline deacylation. Amendment of plant growth substratum with chitosan suppressed the root rot caused by Fusarium oxysporum f. sp. radicis-lycopersici in tomato. Although the exact mechanism of action of chitosan is not fully understood, it has been observed that treatment with chitosan increased resistance to pathogens.

Other microbial byproducts also may contribute to pathogen suppression. Hydrogen cyanide (HCN) effectively blocks the cytochrome oxidase pathway and is highly toxic to all aerobic microorganisms at picomolar concentrations. The production of HCN by certain fluorescent pseudomonads is believed to be involved in the suppression of root pathogens. P. fluorescens CHAo produces antibiotics, siderophores and HCN, but suppression of black rot of tobacco caused by Thielaviopsis basicola appeared to be due primarily to HCN production. Howell et al. reported that volatile compounds such as ammonia produced by Enterobacter cloacae were involved in the suppression of Pythium ultimum-induced damping-off of cotton. While it is clear that biocontrol microbes can release many different compounds into their surrounding environment, the types and amounts produced in natural systems in the presence and absence of plant disease have not been well documented and this remains a frontier for discovery.

Competition

From a microbial perspective, soils and living plant surfaces are frequently nutrient limited environments. To successfully colonize the phytosphere, a microbe must effectively compete for the available nutrients. On plant surfaces, host-supplied nutrients include exudates, leachates, or senesced tissue. Additionally, nutrients can be obtained from waste products of other organisms such as insects (e.g. aphid honeydew on leaf surface) and the soil. While difficult to prove directly, much indirect evidence suggests that competition between pathogens and non-pathogens for nutrient resources is important for limiting disease incidence and severity. In general, soilborne pathogens, such as species of Fusarium and Pythium, that infect through mycelial contact are more susceptible to competition from other soil- and plant-associated microbes than those pathogens that germinate directly on plant surfaces and infect through appressoria and infection pegs. Genetic work of Anderson et al. revealed that production of a particular plant glycoprotein called agglutinin was correlated with potential of P. putida to colonize the root system. P. putida mutants deficient in this ability exhibited reduced capacity to colonize the rhizosphere and a corresponding reduction in Fusarium wilt suppression in cucumber. The most abundant nonpathogenic plant-associated microbes are generally thought to protect the plant by rapid colonization and thereby exhausting the limited available substrates so that none are available for pathogens to grow. For example, effective catabolism of nutrients in the spermosphere has been identified as a mechanism contributing to the suppression of Pythium ultimum by Enterobacter cloacae. At the same time, these microbes produce metabolites that suppress pathogens. These microbes colonize the sites where water and carbon-containing nutrients are most readily available, such as exit points of secondary roots, damaged epidermal cells, and nectaries and utilize the root mucilage.

Biocontrol based on competition for rare but essential micronutrients, such as iron, has also been examined. Iron is extremely limited in the rhizosphere, depending on soil pH. In highly oxidized and aerated soil, iron is present in ferric form, which is insoluble in water (pH 7.4) and the concentration may be as low as 10-18 M. This concentration is too low to support the growth of microorganisms, which generally need concentrations approaching 10-6 M. To survive in such an environment, organisms were found to secrete iron-binding ligands called siderophores having high affinity to sequester iron from the micro-environment. Almost all microorganisms produce siderophores, of either the catechol type or hydroxamate type. Kloepper et al. were the first to demonstrate the importance of siderophore production as a mechanism of biological control of Erwinia carotovora by several plant-growth-promoting Pseudomonas fluorescens strains A1, BK1, TL3B1 and B10. And, a direct correlation was established in vitro between siderophore synthesis in fluorescent pseudomonads and their capacity to inhibit germination of chlamydospores of F. oxysporum. As with the antibiotics, mutants incapable of producing some siderophores, such as pyoverdine, were reduced in their capacity to suppress different plant pathogens. The increased efficiency in iron uptake of the commensal microorganisms is thought to be a contributing factor to their ability to aggressively colonize plant roots and an aid to the displacement of the deleterious organisms from potential sites of infection.

Induction of Host Resistance

Plants actively respond to a variety of environmental stimuli, including gravity, light, temperature, physical stress, water and nutrient availability. Plants also respond to a variety of chemical stimuli produced by soil- and plant-associated microbes. Such stimuli can either induce or condition plant host defenses through biochemical changes that enhance resistance against subsequent infection by a variety of pathogens. Induction of host defenses can be local and/or systemic in nature, depending on the type, source, and amount of stimuli. Recently, phytopathologists have begun to characterize the determinants and pathways of induced resistance stimulated by biological control agents and other non-pathogenic microbes. The first of these pathways, termed systemic acquired resistance (SAR), is mediated by salicylic acid (SA), a compound which is frequently produced following pathogen infection and typically leads to the expression of pathogenesis-related (PR) proteins. These PR proteins include a variety of enzymes some of which may act directly to lyse invading cells, reinforce cell wall boundaries to resist infections, or induce localized cell death. A second phenotype, first referred to as induced systemic resistance (ISR), is mediated by jasmonic acid (JA) and/or ethylene, which are produced following applications of some nonpathogenic rhizobacteria. Interestingly, the SA- and JA- dependent defense pathways can be mutually antagonistic, and some bacterial pathogens take advantage of this to overcome the SAR. For example, pathogenic strains of Pseudomonassyringae produce coronatine, which is similar to JA, to overcome the SA-mediated pathway. Because the various host-resistance pathways can be activated to varying degrees by different microbes and insect feeding, it is plausible that multiple stimuli are constantly being received and processed by the plant. Thus, the magnitude and duration of host defense induction will likely vary over time. Only if induction can be controlled, i.e. by overwhelming or synergistically interacting with endogenous signals, will host resistance be increased.

Table: Bacterial determinants and types of host resistance induced by biocontrol agents.

Bacterial strain	Plant species	Bacterial determinant	Type
Bacillus mycoides strain Bac J	Sugar beet	Peroxidase, chitinase and β-1,3-glucanase	ISR
Bacillus subtilis GB03 and IN937a	Arabidopsis	2,3-butanediol	ISR
Pseudomonas fluorescens strains			
CHA0	Tobacco	Siderophore	SAR
	Arabidopsis	Antibiotics (DAPG)	ISR
WCS374	Radish	Lipopolysaccharide	ISR
		Siderophore	
		Iron regulated factor	
WCS417	Carnation	Lipopolysaccharide	ISR
	Radish	Lipopolysaccharide	ISR
		Iron regulated factor	
	Arabidopsis	Lipopolysaccharide	ISR
	Tomato	Lipopolysaccharide	ISR
Pseudomonas putida strains	Arabidopsis	Lipopolysaccharide	ISR

WCS 358	Arabidopsis	Lipopolysaccharide	ISR
		Siderophore	ISR
BTP1	Bean	Z,3-hexenal	ISR
Serratia marcescens 90-166	Cucumber	Siderophore	ISR

A number of strains of root-colonizing microbes have been identified as potential elicitors of plant host defenses. Some biocontrol strains of Pseudomonas sp. and Trichoderma sp. are known to strongly induce plant host defenses. In several instances, inoculations with plant-growth-promoting rhizobacteria (PGPR) were effective in controlling multiple diseases caused by different pathogens, including anthracnose (Colletotrichum lagenarium), angular leaf spot (Pseudomonas syringae pv. lachrymans and bacterial wilt (Erwinia tracheiphila). A number of chemical elicitors of SAR and ISR may be produced by the PGPR strains upon inoculation, including salicylic acid, siderophore, lipopolysaccharides, and 2,3-butanediol, and other volatile substances. Again, there may be multiple functions to such molecules blurring the lines between direct and indirect antagonisms. More generally, a substantial number of microbial products have been identified as elicitors of host defenses, indicating that host defenses are likely stimulated continually over the course of a plant's lifecycle. Excluding the components directly related to pathogenesis, these inducers include lipopolysaccharides and flagellin from Gram-negative bacteria; cold shock proteins of diverse bacteria; transglutaminase, elicitins, and β-glucans in Oomycetes; invertase in yeast; chitin and ergosterol in all fungi; and xylanase in Trichoderma. These data suggest that plants would detect the composition of their plant-associated microbial communities and respond to changes in the abundance, types, and localization of many different signals. The importance of such interactions is indicated by the fact that further induction of host resistance pathways, by chemical and microbiological inducers, is not always effective at improving plant health or productivity in the field.

Microbial Diversity and Disease Suppression

Plants are surrounded by diverse types of mesofauna and microbial organisms, some of which can contribute to biological control of plant diseases. Microbes that contribute most to disease control are most likely those that could be classified competitive saprophytes, facultative plant symbionts and facultative hyperparasites. These can generally survive on dead plant material, but they are able to colonize and express biocontrol activities while growing on plant tissues. A few, like avirulent Fusarium oxysporum and binucleate Rhizoctonia-like fungi, are phylogenetically very similar to plant pathogens but lack active virulence determinants for many of the plant hosts from which they can be recovered. Others, like Pythium oligandrum are currently classified as distinct species. However, most are phylogenetically distinct from pathogens and, most often, they are subspecies variants of the same microbial groups.

Due to the ease with which they can be cultured, most biocontrol research has focused on a limited number of bacterial (Bacillus, Burkholderia, Lysobacter, Pantoea, Pseudomonas, and Streptomyces) and fungal (Ampelomyces, Coniothyrium, Dactylella, Gliocladium, Paecilomyces, and Trichoderma) genera. Still, other microbes that are more recalcitrant to in vitro culturing have been intensively studied. These include mycorrhizal fungi, e.g. Pisolithus and Glomus spp. that can limit subsequent infections, and some hyperparasites of plant pathogens, e.g. Pasteuria penetrans which attack root-knot nematodes. Because multiple infections can and do take place in field-grown plants, weakly virulent pathogens can contribute to the suppression of more virulent pathogens, via the induction of host defenses. Lastly, there are the many general micro- and meso-fauna predators, such as protists, collembola, mites, nematodes, annelids, and insect larvae whose activities can reduce pathogen biomass, but may also facilitate infection and/or stimulate plant host defenses by virtue of their own herbivorous activities.

While various epiphytes and endophytes may contribute to biological control, the ubiquity of mycorrhizae deserves special consideration. Mycorrhizae are formed as the result of mutualist symbioses between fungi and plants and occur on most plant species. Because they are formed early in the development of the plants, they represent nearly ubiquitous root colonists that assist plants with the uptake of nutrients (especially phosphorus and micronutrients). The vesicular arbuscular mycorrhizal fungi (VAM, also known as arbuscular mycorrhizal or endomycorrhizal fungi) are all members of the zygomycota and the current classification contains one order, the Glomales, encompassing six genera into which 149 species have been classified. Arbuscular mycorrhizae involve aseptate fungi and are named for characteristic structures like arbuscles and vesicles found in the root cortex. Arbuscules start to form by repeated dichotomous branching of fungal hyphae approximately two days after root penetration inside the root cortical cell. Arbuscules are believed to be the site of communication between the host and the fungus.

Vesicles are basically hyphal swellings in the root cortex that contain lipids and cytoplasm and act as storage organ of VAM. These structures may present intra- and intercellular and can often develop thick walls in older roots. These thick walled structures may function as propagules. During colonization, VAM fungi can prevent root infections by reducing the access sites and stimulating host defense. VAM fungi have been found to reduce the incidence of root-knot nematode. Various mechanisms also allow VAM fungi to increase a plant's stress tolerance. This includes the intricate network of fungal hyphae around the roots which block pathogen infections. Inoculation of apple-tree seedlings with the VAM fungi Glomus fasciculatum and G. macrocarpum suppressed apple replant disease caused by phytotoxic myxomycetes. VAM fungi protect the host plant against root-infecting pathogenic bacteria. The damage due to Pseudomonas syringae on tomato may be significantly reduced when the plants are well colonized by mycorrhizae. The mechanisms involved in these interactions include physical protection, chemical interactions and indirect effects. The other mechanisms employed by VAM fungi to indirectly suppress plant pathogens include enhanced nutrition to plants; morphological changes in the root by increased lignification; changes in the chemical composition of the plant tissues like antifungal chitinase, isoflavonoids, etc.; alleviation of abiotic stress and changes in the microbial composition in the mycorrhizosphere. In contrast to VAM fungi, ectomycorrhizae proliferate outside the root surface and form a sheath around the root by the combination of mass of root and hyphae called a mantle. Disease protection by ectomycorrhizal fungi may involve multiple mechanisms including antibiosis, synthesis of fungistatic compounds by plant roots in response to mycorrhizal infection and a physical barrier of the fungal mantle around the plant root. Ectomycorrhizal fungi like Paxillus involutus effectively controlled root rot caused by Fusarium oxysporum and Fusarium moniliforme in red pine. Inoculation of sand pine with Pisolithus tinctorius, another ectomycorrhizal fungus, controlled disease caused by Phytophthora cinnamomi.

Because plant diseases may be suppressed by the activities of one or more plant-associated microbes, researchers have attempted to characterize the organisms involved in biological control. Historically, this has been done primarily through isolation, characterization, and application of individual organisms. By design, this approach focuses on specific forms of disease suppression. Specific suppression results from the activities of one or just a few microbial antagonists. This type of suppression is thought to be occurring when inoculation of a biocontrol agent results in substantial levels of disease suppressiveness. Its occurrence in natural systems may also occur from time to time. For example, the introduction of Pseudomonas fluorescens that produce the antibiotic 2,4-diacetylphloroglucinol can result in the suppression of various soilborne pathogens. However, specific agents must compete with other soil- and root-associated microbes to survive, propagate, and express their antagonistic potential during those times when the targeted pathogens pose an active threat to plant health. In contrast, general suppression is more frequently invoked to explain the reduced incidence or severity of plant diseases because the activities of multiple organisms can contribute to a reduction in disease pressure. High soil organic matter supports a large and diverse mass of microbes resulting in the availability of fewer ecological niches for which a pathogen competes. The extent of general suppression will vary substantially depending on the quantity and quality of organic matter present in a soil. Functional redundancy within different microbial communities allows for rapid depletion of the available soil nutrient pool under a large variety of conditions, before the pathogens can utilize them to proliferate and cause disease. For example, diverse seed-colonizing bacteria can consume nutrients that are released into the soil during germination thereby suppressing pathogen germination and growth. Manipulation of agricultural systems, through additions of composts, green manures and cover crops is aimed at improving endogenous levels of general suppression.

Biocontrol Research, Development and Adoption

Biological control really developed as an academic discipline during the 1970s and is now a mature science supported in both the public and private sector. In the United

States, research funds for the discipline are provided primarily by several USDA programs. These include the Section 406 programs, regional IPM grants, Integrated Organic Program, IR-4, and several programs funded as part of the National Research Initiative. Monies also exist to stimulate the development of commercial ventures through the small business innovation research (SBIR) programs. Such ventures are intended to be conduits for academic research that can be used to develop new companies.

Much has been learned from the biological control research conducted over the past forty years. But, in addition to learning the lessons of the past, biocontrol researchers need to look forward to define new and different questions, the answers to which will help facilitate new biocontrol technologies and applications. Currently, fundamental advances in computing, molecular biology, analytical chemistry, and statistics have led to new research aimed at characterizing the structure and functions of biocontrol agents, pathogens, and host plants at the molecular, cellular, organismal, and ecological levels. Some of the research questions that will advance our understanding of biological controls and the conditions under which it can be most fruitfully applied are liste.

Some current topics of biocontrol research and development and associated questions:

1. The ecology of plant-associated microbes:

   • How are pathogens and their antagonists distributed in the environment?

   • Under what conditions do biocontrol agents exert their suppressive capacities?

   • How do native and introduced populations respond to different management practices?

   • What determines successful colonization and expression of biocontrol traits?

   • What are the components and dynamics of plant host defense induction?

2. Application of current strains/inoculant strategies:

   • Can more effective strains or strain variants be found for current applications?

   • Will genetic engineering of microbes and plants be useful for enhancing biocontrol?

   • How can formulations be used to enhance activities of known biocontrol agents?

3. Discovering novel strains and mechanisms of action:

   • Can previously uncharacterized microbes act as biological control agents?

   • What other genes and gene products are involved in pathogen suppression?

- Which novel strain combinations work more effectively than individual agents?

- Which signal molecules of plant and microbial origin regulate the expression of biocontrol traits by different agents?

4. Practical integration into agricultural systems:

- Which production systems can most benefit from biocontrol for disease management?

- Which biocontrol strategies best fit with other IPM system components?

- Can effective biocontrol-cultivar combinations be developed by plant breeders?

Over the past fifty years, academic research has led to the development of a small but vital commercial sector that produces a number of biocontrol products. The current status of commercialization of biological control products has been reviewed recently. As in most industries, funding in the private sector research and development goes through cycles, but seems likely to increase in the years ahead as regulatory and price pressures for agrochemical inputs increase. Most of the commercial production of biological control agents is handled by relatively small companies, such as Agraquest, BioWorks, Novozymes, Prophyta, Kemira Agro. Occasionally, such companies are absorbed by or act as subsidiaries of multi-billion dollar agrochemical companies, such as Bayer, Monsanto, Syngenta, and Sumitomo. Total revenues of products used for biocontrol of plant diseases represented just a small fraction of the total pesticide market during the first few years of the 21st century with total sales on the order of $10 to 20 million dollars annually. However, significant expansion is expected over the next 10 years due to increasing petroleum prices, the expanded demand for organic food, and increased demand for "safer" pesticides in agriculture, forestry, and urban landscapes.

Growers are interested in reducing dependence on chemical inputs, so biological controls (defined in the narrow sense) can be expected to play an important role in Integrated Pest Management (IPM) systems. A model describing the several steps required for a successful IPM has been developed. In this model, good cultural practices, including appropriate site selection, crop rotations, tillage, fertility and water management, provide the foundation for successful pest management by providing a fertile growing environment for the crop. The use of pest- and disease-resistant cultivars, developed through conventional breeding or genetic engineering, provides the next line of defense. However, such measures are not always sufficient to be productive or economically sustainable. In such cases, the next step would be to deploy biorational controls of insect pests and diseases These include BCAs, introduced as inoculants or amendments, as well as active ingredients directly derived from natural origins and having a low impact on the environment and non-target organisms. If these foundational options are not sufficient to ensure plant health and/or economically sustainable production, then less specific and more harmful synthetic chemical toxins can be used to ensure productivity and profitability. With the growing interest in reducing chemical inputs, companies involved in the manufacturing and marketing of BCAs should experience continued growth. However, stringent quality control measures must be adopted so that farmers get quality products. New, more effective and stable formulations also will need to be developed.

Most pathogens will be susceptible to one or more biocontrol strategies, but practical implementation on a commercial scale has been constrained by a number of factors. Cost, convenience, efficacy, and reliability of biological controls are important considerations, but only in relation to the alternative disease control strategies. Cultural practices (e.g. good sanitation, soil preparation, and water management) and host resistance can go a long way towards controlling many diseases, so biocontrol should be applied only when such agronomic practices are insufficient for effective disease control. As long as petroleum is cheap and abundant, the cost and convenience of chemical pesticides will be difficult to surpass. However, if the infection court or target pathogen can be effectively colonized using inoculation, the ability of the living organism to reproduce could greatly reduce application costs. In general, though, regulatory and cultural concerns about the health and safety of specific classes of pesticides are the primary economic drivers promoting the adoption of biological control strategies in urban and rural landscapes. Self-perpetuating biological controls (e.g. hypovirulence of the chestnut blight pathogen) are also needed for control of diseases in forested and rangeland ecosystems where high application rates over larger land areas are not economically-feasible. In terms of efficacy and reliability, the greatest successes in biological control have been achieved in situations where environmental conditions are most controlled or predictable and where biocontrol agents can preemptively colonize the infection court. Monocyclic, soilborne and post-harvest diseases have been controlled effectively by biological control agents that act as bioprotectants (i.e. preventing infections). Specific applications for high value crops targeting specific diseases (e.g. fireblight, downy mildew, and several nematode diseases) have also been adopted. As research unravels the various conditions needed for successful biocontrol of different diseases, the adoption of BCAs in IPM systems is bound to increase in the years ahead.

Pathogen Evolution across the Agro-ecological Interface

Increased human impacts on all levels of biological organization (e.g. fragmentation of natural systems and changing patterns of landuse, global movement of species), and in many novel ways (e.g. genetically modified organisms, introduction of new resistance genes into crops) has led to the explicit recognition of the value of an applied science of coevolutionary biology in management and planning contexts. Ultimately, the development of such a discipline will increase our ability to predict the outcomes of different types of human intervention (e.g. antibiotics, species introductions). Furthermore, our ability to both disrupt natural coevolutionary interactions as well as shift them in new and as yet unpredictable directions (e.g. emerging diseases) emphasizes the need for this predictive capability.

One arena where coevolutionary principles can be broadly applied lies in understanding dynamical processes and feedbacks across the interface between agricultural production systems and natural plant communities. Increasingly, as agricultural production intensifies or is altered to meet new demands (e.g. environmental forestry) in response to a range of external drivers, there are corresponding changes in the potential for contact with native elements in these landscapes. Of particular interest in this context are the ecological and evolutionary dynamics of plant pathogens that may move across this interface to varying degrees. A key question is the extent to which interactions between agriculture and native ecosystems may alter disease epidemiology and the potential for emergence of new plant pathogens. More generally, host co-infection by different pathogens and conversely the infection of different hosts by a given pathogen may be especially prevalent in agro-ecosystems, leading to novel coevolutionary dynamics.

The temporal and spatial nature of interactions between host plants and their fungal pathogens are forever shifting. Heterogeneity in environmental conditions across the distributional range of pathogens affects their ability to thrive and hence the size, frequency and severity of the impacts they have on their hosts. Against this backdrop though, the interactions between most hosts and their pathogens have an added level of subtle complexity. As they respectively place strong selection pressure on each other this can affect the frequency of resistance and avirulence genes within and among different populations, and indeed aspects of their respective life histories.

That pathogens evolve in response to changes in their hosts has been broadly documented in all types of interactions including human health, veterinary and wildlife diseases, plant-based agriculture and natural communities. Of these, changes in pathogen population structure following the deployment of novel cultivars with new resistance combinations are probably among the best examples of evolution in action. Indeed, the sequential deployment of single resistance genes in wheat in the 1950s and 1960s to counter stem rust, and the extremely rapid response of increased virulence in the previously avirulent pathogen population was summarized in the phrase 'man-guided evolution of the rusts'. However, this phenomenon has not been restricted to this particular interaction, occurring instead in virtually all interactions in which single major genes for resistance have been deployed in crops grown under high density and over large areas (e.g. all rusts and mildew of all cereals, mildews of lettuce).

In some cases, interactions between agricultural crops and their pathogens are largely self-contained (e.g. cereal smuts); in other cases though, crop volunteers (self-sown individual crop plants), wild weedy relatives or even unrelated native plants that are host to the same pathogen may play a significant role in shaping the dynamics and direction of the interaction (e.g. barley scald on volunteer plants and weedy Hordeum species; wheat rust parasitizing alternate hosts). What criteria are important in determining involvement in agricultural pathogen evolution by noncrop plants?; what implications do such interactions have for disease management in agro-ecosystems?; and how can we influence those outcomes.

Spatial and temporal aspects of host and pathogen life-history characteristics are crucial determinants of whether or not hosts and pathogens interact, and if they do, the extent and nature of those interactions. The magnitude and extent of influence wielded by noncrop components on disease will strongly depend on such factors. For example, spatially explicit simulation models have shown that the level of pathogen virulence which most strongly reduces host population size is dependent on host longevity, as well as whether pathogen impacts are felt through reduced fecundity or increased mortality. Other life-history features of the interaction such as the relative spatial scales of host and pathogen dispersal can also profoundly affect patterns of disease incidence and prevalence, as well as the dynamical behaviour of the system.

From an evolutionary perspective, among-population connectivity (as determined by both population structure and dispersal) influences the evolution and maintenance of genetic diversity in host resistance and pathogen virulence genes. Importantly, the relative migration rates of hosts and pathogens play a key role in determining patterns of local adaptation, which can in turn influence spatial patterns of disease. Relatively little work has been performed on the consequences of host specificity, although clearly, plant pathogens vary from those that are highly host specific (e.g. many rusts) through to those that are able to attack a broad diversity of host species (e.g. many soil-borne pathogens like Rhizoctonia). This may be of particular importance in the context of host–pathogen interactions across the agro-ecological interface. For example, recent theoretical work suggests that variation in both the intensity and direction of selection pressures across different host species can strongly impact on the evolution of pathogen transmission, and patterns of host exploitation. In turn this feeds back on the epidemiological behaviour of pathogens.

That crop and noncrop species (whether the latter is an introduced weed or native plant) are attacked by the same species of pathogen or pathogens is, a necessary, but not sufficient, condition for crop pathogen evolution to be significantly affected by the presence of a weedy host. It is the relative phenology of pathogen development on crop and noncrop that is important in determining whether the two cohorts are indeed just different parts of the one population or whether they are essentially isolated from each other despite spatial proximity.

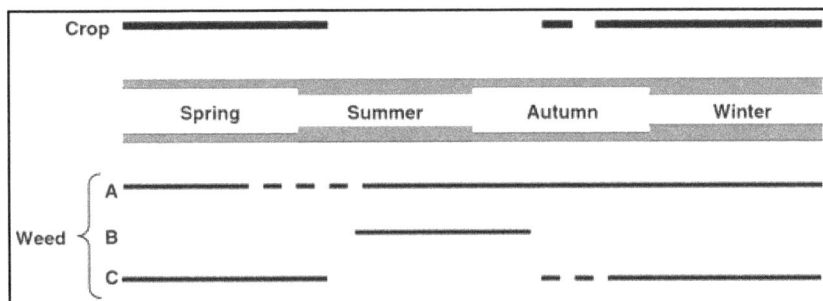

Hypothetical phenological cycles of a crop and associated weeds that
are host to the same pathogens.

Wild or weedy species that act as a 'reservoir' of inoculum providing a 'green bridge' between the maturity of one crop and the appearance of susceptible material in the next, certainly may prevent populations of obligate biotrophic fungi (e.g. rusts and mildews) with no effective resting stage from crashing precipitously with a consequent potential loss of variation through genetic drift or even local extinction. Thus in Australia, weedy Hordeum species have been implicated together with volunteer wheat plants as off-season (over-summering) hosts for Puccinia striiformis allowing this pathogen, that has no protected spore stage, to survive for the several months between maturity of one year's wheat crop and emergence of the next. Similarly, weedy species often act as alternative hosts for many soil pathogens, maintaining pathogen populations at high levels in the absence of the crop host. A good example of this effect is seen in the importance of weedy species in contributing to levels of Fusarium spp. in agricultural soil. Recognition of this potential impact is at the heart of many crop sanitization programmes. Finally, weedy species that are only present at the same time as the crop are unlikely to have a reservoir effect although they may still play an important evolutionary role in affecting the genetic structure of the pathogen population.

Genetic Interactions

While epidemiological impacts may be the most obvious way in which crops and non-crop species affect one another, the most significant and far-reaching effects flow from more discreet interactions that generate genetic change in pathogen populations. What is it about the agro-ecological interface that makes it potentially such a hot-bed for evolutionary change? This boundary is essentially a breakpoint in a continuum in a broad range of ecological, environmental and species diversity measures. On the one hand, are crop communities that, notwithstanding the use of mixtures, alley cropping and other mixed approaches in horticulture, are essentially high density – low diversity and genetically controlled by man's direct intervention. On the other hand, nearly all natural or even semi-natural plant communities are characterized by considerable inter and intra-specific diversity and lower individual species densities. Adding to this complexity, in many cases noncrop plant communities are composed of mixtures of native and exotic plant species, the latter ranging from introduced weeds with low diversity to those that are quite diverse (e.g. due to multiple introductions). These will vary in the suite of life histories represented and the degree of susceptibility to pathogen attack (with direct consequences for pathogen evolution).

Moreover, the potential for contact between agricultural and noncrop communities will itself vary across landscapes depending on the extent and configuration of cropping systems. For example, in the Middle East, there are a broad range of wild Cicer species which grow in varying degrees of proximity to cultivated chickpea (Cicer arietinum) and other crop legumes. Recent work suggests that the occurrence and severity of ascochyta diseases on the crops may be related to the proximity of wild hosts.

However, this is not to imply that evolution only or even predominantly, occurs in the agro-ecological zone. There are many examples of evolutionary change occurring within both agricultural and natural ecosystems. Evolutionary changes in pathogen populations in natural systems have often been assumed rather than demonstrated but clear examples exist of the appearance of novel pathotypes in populations (migration), year-to-year fluctuations in pathotype frequency (drift), changes within seasons (selection) and even much more significant changes resulting from genetic recombination between distinctly different genetic linkages or different species.

Evolutionary change occurring in pathogen populations attacking agricultural crops is also a well-recognized phenomenon but these changes typically reflect within-species changes – the evolution of novel virulence in response to the deployment of crop varieties carrying new resistance genes (mutation) or the re-assortment of characters through sexual recombination or the merging of closely related lineages. Not surprisingly, given the highly selected nature of most crops, the opportunities for major recombination between distinctly different but related pathogen species is very limited as they are rarely found on the one host.

The agro-ecological interface sits like a 'hybrid' zone between these two extremes, and like a hybrid zone is a site of overlap where crop species (whether cereals, horticultural crops or trees) will be in closer proximity to other related species (whether weedy or wild) than elsewhere, and where pathogens of these crops and related species (and hence more closely related pathogens) are also in closer contact. This situation is ripe with opportunities for pathogen evolution as a consequence of an increased diversity of selective pressures – different host species with potentially different resistance genes or mechanisms; novel encounters by related pathogen species. Indeed, in the agro-ecological zone the full gamut of potential changes to pathogens may occur, single virulence mutations to massive whole genome recombination, immediately adjacent to large uniform testing grounds – the agricultural crop! It is worth noting that such situations are not unique to plant–pathogen interactions. In the human context, similar 'testbeds for disease' exist in hospitals which are well-known hotspots for disease emergence and evolution. Basically, any situation which brings together variation (and enhances disease transmission potential) will clearly provide more opportunities for evolution.

Sources of Variation

As with all other species, mutation and recombination are the ultimate source of genetically based variation in pathogen populations. Selection, random drift and migration (gene-flow) then acts on this basic variation to shape the structure of individual populations and hence the way they interact with their environment. Pathogen populations in agricultural crops experience a substantially different selective environment to those in natural ecosystems. The large and genetically uniform plant populations of agriculture provide a relatively uniform/predictable selective environment. Furthermore, notwithstanding the deployment of multiple varieties in the same general area, their use in large uniform blocks ensures that the fraction of the pathogen population that moves from one selective environment to another (e.g. one variety to another) is lower than in comparable wild situations where small, mixed host populations with multiple resistance gene combinations are the norm rather than the exception. The overall numerical and genetical dynamics of pathogen populations at the agro-ecological interface are therefore much better viewed as complex multi-population and multi-species interactions (i.e. metapopulations or even metacommunities) in which environmental, temporal and spatial heterogeneity, and marked differences in life-history features of hosts and pathogens may exert significant evolutionary influence.

Mutation – The Wild Host as a Selective Agent

Random mutation to virulence occurs in all pathogen populations. However, the probability that any such mutation amounts to any more than an ephemeral spark of evolution is dependent on a number of factors, not least of which will be any selective advantage such a mutation may gain through an ability to overcome resistances within the host population. In a genetically uniform crop pathogen population the appearance of a novel virulent pathotype capable of overcoming the deployed resistance may easily lead to a selective sweep culminating in the presence of a single clone or clonal lineage (this is particularly the case in asexual foliar pathogen populations). In diverse host populations, however, such a loss of diversity is far less likely to occur as different hosts with different resistance genes exert different selection pressures on the pathogen population. While it has been argued that over time repeated selection on hosts with different resistance gene combinations would lead to dominance by a single 'super-race', in reality in both agricultural mixtures and natural systems this tendency appears to be counter-balanced by greater fitness (increased fecundity) of isolates with lower virulence. A potential consequence of these interactions is then, the possibility that genetically diverse weedy or wild host populations existing at the agro-ecological interface may be effective generators of pathogen diversity.

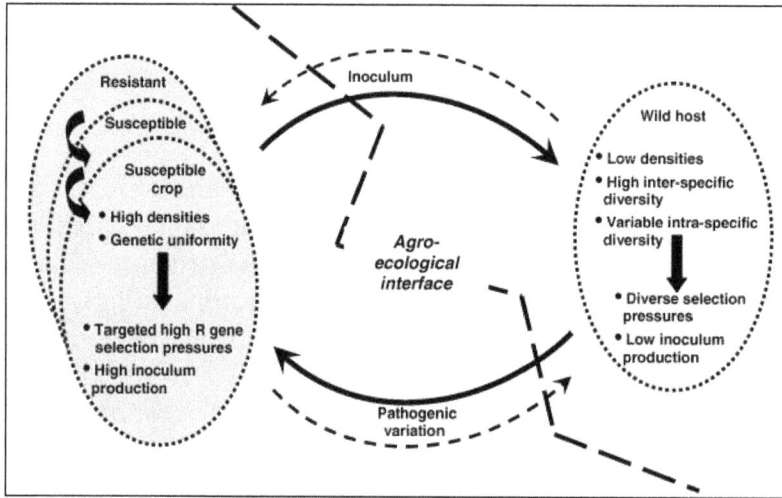

Interactions at the agro-ecological interface between crops and wild and weedy
plants leading to an interchange of inoculum and pathogen variation.

Although there are good examples of no resistance being present, intra and inter-popula-
tion diversity in resistance phenotypes is the norm rather than the exception in wild and
weedy plant species that are hosts of agricultural pathogens (for example, resistance to
Rhynchosporium secalis in Hordeum leporinum, to Puccinia graminis avenae and Puccinia coronata in Avena spp., and to Bremia lactucae in Lactuca spp.. However, the extent of
such variation will be affected by a range of issues surrounding the history of the species
involved. Thus in the case of recent weedy immigrants to a new environment, the number
and magnitude of introductions, the length of time that has elapsed since such introductions, and the disease pressure experienced in the new habitat will all affect the extent of
resistance gene diversity. Examples of the impact of these factors in weedy crop relatives
that are host to agricultural pathogens have not been documented, but the marked difference in diversity of resistance found in Chondrilla juncea to its rust pathogen, Puccinia
chondrillina, between populations in its native range and in Australia illustrates this effect.

In all these cases, the 'matching' crop species typically occurs in large uniform stands in
which resistance diversity within individual stands is very limited. Unfortunately, while
screening plant populations for resistance patterns may be slow, assessing the virulence structure of matching pathogen populations is extremely laborious and has been
performed very infrequently. As a consequence, evidence available for the agro-ecological interface as a major source of novel pathogen phenotypes is largely circumstantial.
However, the most parsimonious explanation for the occurrence of increased patho-
typic diversity found on wild and weedy plant populations at the A–E interface is selection on the range of resistance phenotypes found in those diverse host populations.

Gene flow between Pathogens of Weeds and Crops

The interaction occurring between oats both cultivated (Avena sativa) and wild (Avena
barbata, Avena fatua) and its stem and crown rust pathogens (P. graminis f.sp. avenae

and P. coronata respectively) in Australia provides good circumstantial evidence for the role that wild hosts play in both selectively favouring novel mutations as well as contributing variation to the pathogen population on a cultivated crop through gene flow. In Australia, the sexual hosts of both of these pathogens are absent and although a number of different cultivated oat varieties are typically grown in the same geographic region, each cultivar is essentially a uniform line. In contrast, the wild oat populations show significant diversity in their resistance. Thus, studies of the occurrence of resistance to these two rust pathogens in 21 wild oat populations in eastern Australia found marked differences in the frequency of different seedling resistance phenotypes within and among different host populations. Furthermore, within individual wild oat populations differences were found in both the response of individual host lines to different pathotypes of either pathogen, and between different host lines in their response to any one pathotype. This stands in clear contrast to the pattern of resistance in individual cultivated oat populations where essentially all individuals have the same resistance phenotype.

Against this backdrop of marked differences in resistance, comparisons of the diversity and virulence structure of pathogen populations occurring on either cultivated or wild oat populations from the same areas found no significant differences. However, spatial differences in the resistance structure of the wild oat populations (southern versus northern populations) was positively correlated with the complexity of the virulence structure of pathogen populations in the area – a result that provides strong circumstantial support for a major role for the wild oat populations in driving virulence evolution in the pathogen populations.

It is also relevant to note here the possibility of gene flow between cultivated and wild host plant populations as this may alter the dynamics of any particular host–pathogen interaction. Naturally occurring introgressive hybridization leading to acquisition of novel genes or alleles in wild plants has been documented for a range of traits including, neutral isozyme markers in cultivated–wild oat associations, herbicide resistance flowing from wheat to jointed goat grass (Aegilops cylindrica) and mildew resistance genes from exotic to native gooseberry species. Where such introgression results in the flow of novel resistance genes introduced by breeders from newly released crop varieties into closely related weedy species this may ultimately reduce the range of disease control options available. For example, disease control strategies based around the controlled spatial or temporal use of resistance genes would become less efficient if such resistances remained in the environment (and hence continued to apply selection pressure) after crop varieties containing those genes were removed.

Recombination in Pathogen Populations

Sexual Recombination

Classic examples of this phenomenon where there is a clear role played by a noncrop host are found amongst heteroecious cereal rust fungi where an indeterminate number of asexual pathogen generations on cereals alternates with a sexual stage occurring on either herbaceous or woody wild species. Thus, the occurrence of barberry (Berberis spp.) or buckthorn (Rhamnus spp.) bushes in hedgerows or wooded areas adjacent to cereal crops has been repeatedly shown to be a source of significant pathogen variation in wheat stem, and oat crown rust. Indeed, it is widely accepted that the barberry eradication campaigns that were waged in the Great Plains of the USA in the first third of the 20th Century were responsible for a significant reduction in the overall pathogenic diversity of the P. graminis tritici population and in the frequency of major stem rust epidemics. In that situation, eradication of the barberry not only reduced the diversity of pathotypes locally available to attack newly deployed wheat varieties, but also very substantially reduced the level of inoculum to which the developing crop was subject early in the spring. The lack of sexual recombination for more than 70 years has resulted in a population composed of a limited number of pathotypes that represent a few distinct clonal lineages. Each lineage is composed of a number of closely related pathotypes with one or two virulence differences, while among lineage differences are typically in the order of eight to 10 virulences. Further evidence for the powerful effect of recombination in generating pathogenic diversity is seen in comparisons of the Great Plains populations with those populations from the Washington/Oregon area where the sexual stage of P. graminis tritici occurs annually. In the latter area, the pathogen population is composed of a very broad array of different pathotypes with no evidence of linkage disequilibrium. Where the alternate host is herbaceous, as occurs in the interaction between wheat, Ornithogalum umbellatum and Puccinia hordei the interaction can be even more spatially intimate with the Ornithogalum sexual host growing within the cereal stand itself.

Somatic Recombination

A feature common too many newly emerging diseases – whether they attack plants or animals (including humans) is their appearance in situations where changes in agricultural practices and human activities have either moved existing pathogens large distances around the world or brought species into new proximity. In the world of fungal plant pathogens, there is a steady increasing number of such examples which provides the basic minimum conditions needed (intimate proximity) for the creation of new species through horizontal gene transfer. In this process, genetic exchange occurring via anastomosis of vegetative hyphae is widely regarded as the most likely mechanism, although, given that direct observation of this process has rarely occurred and the existence of fungal nonself recognition systems, the precise mechanism whereby exchange occurs is still in question. Exchange may result in transfer of nuclear material – whole nuclei, individual chromosomes and or of other cytoplasmic components including plasmids and viruses.

Regardless of the actual mechanism, increasingly, evidence provided by sequence data is providing confirmatory support for claims, based on changed pathogenicity patterns, of horizontal gene transfer in fungi. Indeed, with the growing use of sophisticated molecular markers it has become increasingly apparent that such somatic hybrids are more common than previously thought. For example, a number of novel pathogens of trees of agricultural landscapes have arisen in this way. Thus novel hybrid poplar rusts with a broader host range within Populus spp. has resulted from the transfer of pathogenicity genes between different species of Melampsora infecting poplars while a hybrid species of Phytophthora responsible for the deaths of alder trees in Europe has arisen from horizontal gene transfer between Phytophthora cambivora (a pathogen of hardwood trees) and Phytophthora fragariae (a pathogen of strawberries and raspberries). Although these examples do not necessarily relate to interactions at the agro-ecological interface, they do provide strong evidence for the potential for such events and their possible implications.

An example that truly sits along the 'fence-line' between agricultural crops and wild and weedy plants is the Australian 'scabrum' rust story, where a native wild wheat relative (Agropyrum scabrum) that is host to both P. graminis f.sp. tritici and P. graminis f.sp. secalis (stem rust of wheat and rye respectively) provided a common host on which a somatic hybrid naturally developed. The hybrid rust, generated through the exchange of haploid nuclei between the two dikaryotic parent formae speciales, has a different host range to either parent, having acquired through this recombination process virulence on a range of Hordeum vulgare lines that are immune to either parent.

Implications for Crop Management

The role that weeds and native plants play as reservoirs of crop pathogens influencing the incidence and timing of disease in agricultural crops has been widely recognized. However, even in these seemingly simple numerical interactions not everything is as it might seem. Thus cryptic speciation is common among plant pathogens and one cannot assume that because the same species sensu lato occurs on a crop and on wild species nearby there is necessarily any interaction between them. Morphologically identical 'species' may indeed to be biologically separate entities with no effective genetic exchange.

When consideration is turned to the processes of evolutionary change in plant pathogens and how this may be affected by the spatial, temporal and biological complexities introduced at the agro-ecological interface, it becomes increasingly clear that provided the pathogen populations occurring on crop and wild plants are part of the same metapopulation, even relatively uncommon events occurring off the crop may be of great agricultural consequence. The important message is that plant pathogens are amazingly labile organisms. Here, we have provided a range of examples in which the evolution of those typically regarded as crop pathogens may be altered by interactions occurring on other hosts that are wild or weedy species growing in the same general vicinity.

The example provided by the North American experience with Berberis and its impact on wheat rust epidemics – both through a reservoir effect leading to early epidemic development, but more importantly on the evolutionary flexibility of the pathogen in the face of disease resistance breeding strategies is a classic demonstration of the potential of such interactions. In that example, knowledge of the life cycle of the pathogen, and the role of the sexual stage in generating new variation provided the scientific support for the massive eradication campaigns in the American Mid-West in the 1920s and 1930s. Similarly knowledge of the role of wild oats in generating pathogenic variation in oat crown rust has highlighted the futility of a standard major gene resistance approach to rust in oats if no attempt is made to simultaneously tackle wild oat populations.

Ecological Disease Management

In order for a plant to become diseased three conditions must be met: a pathogen, a favorable environment where the pathogen can thrive, and a susceptible host.

Manage the Pathogen

Exclusion

Keep the pathogens out! Make sure that your seeds and transplants are free of pathogens. The epidemic of late blight in Pennsylvania in 2009 was partially the result of the widespread distribution of the pathogen on infected transplants. Consider growing your own transplants; otherwise, inspect them carefully before you bring them to the farm.

Saving seed can easily carry over pathogens from the previous year. Only save seed from healthy plants to reduce this risk. Although pathogens may occasionally be introduced from a commercial seed source, generally they are the most reliable source.

It is also essential to keep your equipment and stakes clean and sanitized. You don't want to be the grower who has bacterial spot in your peppers every year because you reuse your stakes without sanitizing them. Sodium hypochlorite at 0.5 percent (12 times the dilution of household bleach; note that household bleach contains additives and is not allowed for certified organic) is effective and must be followed by rinsing and proper disposal of solution. Hydrogen peroxide is also effective. Prior to sanitizing, remove visible organic debris from the stakes and/or equipment. Organic matter can quickly neutralize surface disinfectants, rendering them ineffective. Also, for the most effective results, change the solution when it becomes visibly dirty. Freezing does not sanitize stakes and equipment.

These materials are on the list of allowed substances for certified organic production. However, it is important, even for allowed materials, to list them on your organic system plan. Any materials you use on certified farms must be cleared with your certifier before use to prevent mishaps that could result in losing certification.

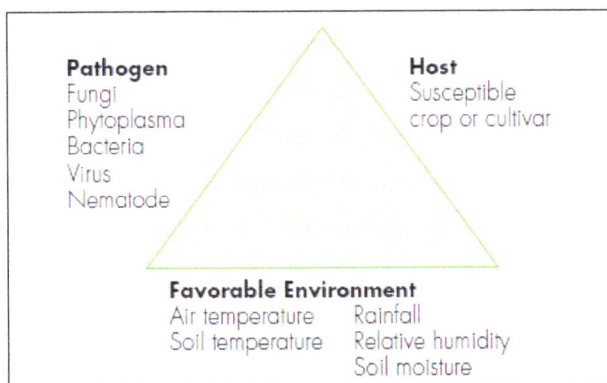

Pathogen
Fungi
Phytoplasma
Bacteria
Virus
Nematode

Host
Susceptible
crop or cultivar

Favorable Environment
Air temperature   Rainfall
Soil temperature   Relative humidity
                     Soil moisture

The Disease Triangle. Rotations that include a fallow period can be the key for controlling some pathogens that have a wide host range.

Eradicate or Reduce the Inoculum

Crop rotation between plant families can help keep the levels of disease down. Rotating to remediate a disease problem can be challenging, especially if the pathogen is long-lived in the soil and/or has a wide host range. Rotating between unrelated crops such as beans to sweet corn, lettuce to cucurbits, and cucurbits to crucifers can help avoid the buildup of soilborne pathogens. For example, northern root-knot nematode is a fairly common problem that attacks carrots and potatoes in addition to a number of other vegetable crops. In a study in New York, when field corn (which is not a host to the nematode) was included in the rotation, the number of nematodes was greatly reduced.

In general, grasses (monocots) are not susceptible to the same diseases as vegetables (dicots). Adding sweet corn, wheat, or a grass cover crop to your rotation can reduce soilborne disease problems. A good rule of thumb is that no crop family should return to the same field or bed for a minimum of three years to avoid soilborne disease buildup. Beware; some pathogens create special structures that allow them to survive in the field for much longer.

Bacterial spot on tomato can spread from debris on reused tomato stakes.

Many pathogens can survive on debris over the winter. Tilling in plant residue at the end of the season allows soil microorganisms to break the material down, leaving potential pathogens without a host.

Antagonistic Plants

Certain plants, such as mustards and sudangrass, can kill plant pathogens that live in the soil. They contain a chemical and an enzyme in their plant tissue, specifically their cell wall. When you mow the plant and crush the tissue, the enzyme reacts with the chemical to create a toxic gas--the same as a fumigant. If you quickly incorporate this crushed plant material into the soil after mowing, the volatilized chemical can kill plant pathogens, nematodes, and weed seeds. Farmers in Northampton County tried this method to control plant-parasitic nematodes.

They learned that the process can be tricky. You need to make sure you have the right varieties, enough moisture, adequate fertility, and good timing to get the result you want.

Hot Water Treatments

If you think your seed might be affected by plant pathogens, you can use a hot water bath that will both surface disinfect as well as kill pathogens within the seed. For example, hot water treatment for eggplant submerses seeds in 122 °F water for 25 minutes. Be careful to find out the correct temperature and length of time for the treatment. Too cold will not work and too hot will kill the seed.

Crop Rotation Affects Pathogen Persistence

Clubroot, caused by Plasmodiophora brassica, can be a significant problem in brassica crops. The pathogen can survive in the soil for over seven years, even in the absence of mustard family crops or weeds. But clubroot tends to decline more quickly when tomato, cucumber, snap bean, and buckwheat are grown.

Clubroot was effectively controlled by growing aromatic herbs including peppermint, garden thyme, and summer savory for two to three consecutive years. For some brassica crops, resistant varieties are also an option.

Create an Unfavorable Environment

Keep Leaves Dry

Most fungi and bacteria that kill plants require wet environments from dew, rain, or irrigation to infect and cause disease. If you want to keep them from reproducing, don't give them the environment they like. Good air circulation and drip irrigation help keep the leaves dry and the diseases out. For example, gray mold in tomatoes is generally not a problem in the field. But, when you pack tomatoes in a high tunnel with little air circulation, it becomes common, especially within the lower portion of the plant.

Maintain High-quality Soil

Balanced fertility, good drainage, and good soil tilth will all help promote a diverse range of soil microorganisms. Diverse microbial communities generally compete with plant-pathogenic organisms in the soil and help keep your plants healthier. Additionally, plants that are not stressed are less susceptible to disease.

Manage Weeds

Many weeds are also hosts for diseases. When your crop is surrounded by weeds, the atmosphere tends to be moist, favoring infection.

Choose a Less Susceptible Variety

Using disease-resistant varieties is one of the most economical and reliable methods of disease management. Resistant varieties are not available for all diseases of vegetable crops, but they definitely should be considered. Dr. McGrath from Cornell University maintains an excellent list of disease resistant cultivars. Your seed catalog will also list disease resistance. Note the letters DM, PM, etc., after each cultivar. They are codes to tell you which diseases the cultivar is resistant to. It is important to become familiar with common vegetable diseases in your region.

Resistant varieties are rarely immune to the disease. They do help delay the onset of disease development, therefore potentially increasing your yields and allowing your crop to fully mature.

How to Hot Water Treat Seed

Step 1: Wrap seeds loosely in a woven cotton (such as cheesecloth) or nylon bag.

Step 2: Prewarm seed for 10 minutes in 100 °F (37 °C) water.

Step 3: Place prewarmed seed in a water bath that will constantly hold the water at the recommended temperature (see below). Length of treatment and temperature of water must be exactly as prescribed.

- Lettuce, celery: 118°F for 30 minutes.

- Broccoli, cauliflower, carrot, collard, kale, kohlrabi, turnip: 122°F for 20 minutes.

- Brussels sprouts, eggplant, spinach, cabbage, tomato: 122 °F for 25 minutes.

- Pepper: 125 °F for 30 minutes.

Step 4: After treatment, place bags in cold tap water for 5 minutes to stop heating action.

Step 5: Spread seed in a single, uniform layer on screen to dry.

Check with your Extension office for possible local hot water seed treatment stations.

Pest Management Materials

Early detection is important for successful disease management. Make sure you scout plants regularly and know which diseases are present in the crop. When preventive and cultural methods for disease control are insufficient to manage a disease, National Organic Program (NOP) compliant inputs can be applied.

Before applying any pest control product, be sure to:

1. Read the label to make sure the product is labeled for the crop and the disease you are trying to manage,

2. Read and understand the safety precautions and application restrictions,

3. Confirm that the brand name product is listed in your Organic System Plan and approved by your certifier.

Permissions

All chapters in this book are published with permission under the Creative Commons Attribution Share Alike License or equivalent. Every chapter published in this book has been scrutinized by our experts. Their significance has been extensively debated. The topics covered herein carry significant information for a comprehensive understanding. They may even be implemented as practical applications or may be referred to as a beginning point for further studies.

We would like to thank the editorial team for lending their expertise to make the book truly unique. They have played a crucial role in the development of this book. Without their invaluable contributions this book wouldn't have been possible. They have made vital efforts to compile up to date information on the varied aspects of this subject to make this book a valuable addition to the collection of many professionals and students.

This book was conceptualized with the vision of imparting up-to-date and integrated information in this field. To ensure the same, a matchless editorial board was set up. Every individual on the board went through rigorous rounds of assessment to prove their worth. After which they invested a large part of their time researching and compiling the most relevant data for our readers.

The editorial board has been involved in producing this book since its inception. They have spent rigorous hours researching and exploring the diverse topics which have resulted in the successful publishing of this book. They have passed on their knowledge of decades through this book. To expedite this challenging task, the publisher supported the team at every step. A small team of assistant editors was also appointed to further simplify the editing procedure and attain best results for the readers.

Apart from the editorial board, the designing team has also invested a significant amount of their time in understanding the subject and creating the most relevant covers. They scrutinized every image to scout for the most suitable representation of the subject and create an appropriate cover for the book.

The publishing team has been an ardent support to the editorial, designing and production team. Their endless efforts to recruit the best for this project, has resulted in the accomplishment of this book. They are a veteran in the field of academics and their pool of knowledge is as vast as their experience in printing. Their expertise and guidance has proved useful at every step. Their uncompromising quality standards have made this book an exceptional effort. Their encouragement from time to time has been an inspiration for everyone.

The publisher and the editorial board hope that this book will prove to be a valuable piece of knowledge for students, practitioners and scholars across the globe.

Index

www.ingramcontent.com/pod-product-compliance
Lightning Source LLC
Chambersburg PA
CBHW061948190326
41458CB00009B/2822

9 781647 400057